Advances in Responsible Land Administration

Advances in Responsible Land Administration

Edited by

Jaap Zevenbergen
Walter De Vries
Rohan Bennett

CRC Press
Taylor & Francis Group
Boca Raton London New York

CRC Press is an imprint of the
Taylor & Francis Group, an **informa** business

CRC Press
Taylor & Francis Group
6000 Broken Sound Parkway NW, Suite 300
Boca Raton, FL 33487-2742

First issued in paperback 2019

ISBN-13: 978-1-4987-1959-9 (hbk)
ISBN-13: 978-0-367-87282-3 (pbk)

Library of Congress Cataloging-in-Publication Data

Advances in responsible land administration / edited by Jaap Zevenbergen, Walter De Vries, and Rohan Mark Bennett.
pages cm
Includes bibliographical references and index.
ISBN 978-1-4987-1959-9 (hardcover : alk. paper) 1. Land use--Management. 2. Land use--Planning. I. Zevenbergen, J. A. (Jaap A.) II. De Vries, Walter, editor of compilation. III. Bennett, Rohan Mark, editor of compilation.

HD111.A28 2016
333.73'13--dc23 2015017779

Visit the Taylor & Francis Web site at
http://www.taylorandfrancis.com

and the CRC Press Web site at
http://www.crcpress.com

Contents

Foreword

Although the positive statistics in the *Millennium Development Goals Report* (United Nations 2013) have instilled in the world a sense of hope, the global challenges of poverty reduction and sustainable development remain significant and daunting. The International Fund for Agricultural Development (2011) points out that despite there being significant progress in reducing poverty—particularly in east Asia—around 1.1 billion people live on less than USD1.25 a day and close to 1 billion people suffer from hunger. It adds that during the 2006–2008 food crises, some 100 million urban and rural poor people joined the ranks of the world's hungry. UN-Habitat (United Nations Human Settlements Programme 2008) reports that the world's slum population is expected to increase to 1.4 billion and urbanization will continue to contribute significantly to climate change issues: the world's 20 largest cities consume 80% of the world's energy, and urban areas generate 80% of the global greenhouse gas emissions (FIG 2010). Although there are various key strategies and interventions to address poverty reduction and, more generally, sustainable development, land governance initiatives are critically important. In this context, the development of responsible land administration systems is strategic and provides one bridge across the divide between those with adequate access to land and those without.

Various authors articulate the potential benefits of land administration systems: poverty alleviation, security of tenure, management of land disputes, inclusive planning, management of natural resources, and environmental protection are all enabled (Antonio 2006; Burns et al. 2007; Magel 2006; Williamson et al. 2010). However, in developing contexts extending existing cadastres, land registries, and land administration systems to those areas not covered by any remains an enormous challenge (Lemmen et al. 2009; Lemmen 2010; UN-Habitat 2008, 2012). Enemark et al. (2014) suggest that 75% of the world's population is not serviced by formal land administration systems that could better safeguard their land rights. The majority are the poor, women, and vulnerable groups, all with limited land access and limited resources to challenge the prevailing situation. These groups experience insecurity of tenure first hand: they have no legally documented or registered land rights and are often forgotten or overlooked during the so-called developmental projects. However, what they do possess are informal, customary, and unwritten land interests, which in most cases are subject to overlapping claims, disputes, and conflict. Land experts recognize that these tenure types cannot be included in conventional land registration systems (United Nations Economic Commission for Africa 1998): parcel-based spatial descriptions of land rights are only one part of a larger and more complex tenure landscape.

It is in this context that the Global Land Tool Network (GLTN) and its more than 65 international partners have explored and developed pro-poor and gender-appropriate land policies and tools. Facilitated by UN-Habitat, the GLTN is a global partnership of international actors working to address global land tenure and land governance challenges. GLTN partners recognize and promote the concept of a continuum of land rights (UN-Habitat 2008, 2012). Along the continuum, different tenure types are featured: the complex interactions between individuals, groups, land resources, plots, dwellings, and settlements should all be catered to. As GLTN partners express it, conventional land titling approaches have largely failed to deliver their expected results because the existing technical solutions are expensive, inappropriate for the diversity of tenures found in developing countries, and unsustainable in terms of finance or available capacity. Individual land titling, by itself, is too slow and cannot deliver security of tenure to the majority of people in the developing world (Zevenbergen et al. 2013).

Although the concept is incrementally and widely accepted in global discourse, a new set of land administration and information management systems are necessary to support and implement the continuum of land rights. In practical terms, this means responding to the needs and requirements of 70% of the citizenry in a developing country that mostly comprises poor people and providing them with equal socioeconomic opportunities. In this regard, the University of Twente's ITC Faculty, Enschede, The Netherlands, where the coeditors of this book are all based, has an excellent track record of contributing to innovative land tenure approaches and supportive land tool designs. The ITC was one of the founding partners of the GLTN and has since assisted in the design of multiple land tools, including the Social Tenure Domain Model (STDM) (see Chapter 15). That the ITC is deeply involving its MSc and PhD students in its land tool research is a testament to the seriousness with which it undertakes its mission. The approach is helping young professionals to build new understandings about land tenure, broaden horizons on design and evaluation, and transfer the knowledge to colleagues and counterparts globally.

When land administration is implemented responsibly, it underpins good land governance and ultimately supports sustainable development. This book is a compilation of lessons about responsible land administration as it has been developed and applied across different parts of Africa, a continent aware of its land challenges and its opportunity to lead the way in the implementation of fit-for-purpose approaches. Writing from Nairobi, Kenya, where we coordinate the GLTN, it is clear that this book will contribute to our work within UN-Habitat and GLTN and also the broader United Nations family: responsible land administration can contribute to more equitable relationships between people and between people and land. Responsible land administration will have its fullest impact when incorporated into wider good land governance frameworks and when implemented as a tool of a well-considered and appropriate land policy. The contributions made by books such as this mean that advances in the latter will no longer be lost due to a lack of appropriate land tools.

Dr. Clarissa Augustinus
Land and GLTN Unit
Urban Legislation, Land and Governance Branch
United Nations Human Settlement Programme
UN-Habitat/GLTN
Nairobi, Kenya

References

Antonio, D. R. 2006. "Instituting good governance in the land administration system: the Philippines' perspective." In *Land Reform, Land Settlement and Cooperatives*. Vol. 2006, p. 73–83. Rome: FAO.

Burns, T., C. Grant, K. Nettle, Anne-Marie Brits, and K. Dalrymple. 2007. "Land administration reform: indicators of success and future challenges." *Agriculture and Rural Development Discussion Paper* 37:1–227.

Enemark, S., K.C. Bell, C.H.J. Lemmen, and R. McLaren. 2014. *Fit-for-Purpose Land Administration: Open Access e-Book*. Vol. 60, FIG publication. Copenhagen, Denmark: International Federation of Surveyors (FIG).

FIG. 2010. "Rapid urbanization and mega cities: the need for spatial information management." In *FIG Publication No. 48*, edited by FIG—International Federation of Surveyors. Copenhagen, Denmark: FIG.

International Fund for Agricultural Development. 2011. *Rural Poverty Report 2011*. Rome: IFAD.

Lemmen, C.H.J. 2010. *Social Tenure Domain Model: A Pro-Poor Land Tool: e-Book*. Vol. 52, FIG publication. Copenhagen, Denmark: International Federation of Surveyors (FIG).

Lemmen, C.H.J., C. Augustinus, and P.J.M. van Oosterom. 2009. "The social tenure domain model." *GIM International* 23 (10):3.

Magel, H. 2006. "Promoting land administration and good governance." A keynote address to the 5th FIG Regional Conference, Accra, Ghana.

UN-Habitat. 2008. *Secure Land Rights for All*. Nairobi, Kenya: UN-Habitat, GLTN.

UN-Habitat. 2012. *Handling Land: Innovative Tools for Land Governance and Secure Tenure*, edited by Global Land Tool Network (GLTN). Nairobi, Kenya: GLTN.

United Nations. 2013. *The Millennium Development Goals Report*. New York.

United Nations Economic Commission for Africa. 1998. *An Integrated Geo-information System (GIS) with Emphasis on Cadastre and Land Information Systems for Decision-Makers in Africa*. Background Report of Expert Group Meeting, Addis Ababa, Ethiopia.

United Nations Human Settlements Programme. 2008. *State of the World's Cities 2010/2011—Cities for All: Bridging the Urban Divide*. Nairobi, Kenya.

Williamson, I.P., S. Enemark, J. Wallace, and A. Rajabifard. 2010. *Land Administration for Sustainable Development*. Redlands, CA: ESRI.

Zevenbergen, J., C. Augustinus, D. Antonio, and R. Bennett. 2013. "Pro-poor land administration: principles for recording the land rights of the underrepresented." *Land Use Policy* 31:595–604. doi: 10.1016/j.landusepol.2012.09.005.

Acknowledgments

We thank all those individuals and organizations that have supported the development of this book. In particular, we acknowledge our colleagues at the Faculty of Geoinformation Science and Earth Observation (ITC) at the University of Twente, Enschede, The Netherlands, for their input and guidance during the conceptualization, review, and editing phases. We would also like to acknowledge our collaborating institutions and host organizations that supported fieldwork, writing activities, and the review process. In addition, we thank each of the authors for their dedication and perseverance in the development of the individual chapters. The contributors of this book are closely tied to professional research projects and, most of all, are experts in the field of land administration. Each of the authors has worked in or with land administration organizations that were at times constrained by limited resources and capacities. This motivated the collective of authors to study, design, implement, and evaluate alternative methods and solutions, ones more aligned toward their own or similar contexts, in other words, toward alternative forms of land administration that are more responsible and fit-for-purpose. The collaborative effort of this book consolidates practice-based longitudinal knowledge and sets new possible directions for future research in land administration, both in methodology and in content.

Editors

Jaap Zevenbergen is a full professor of land administration and management at the University of Twente, Enschede, The Netherlands. He has extensive experience with design and evaluation of recording or registration of land tenure rights, legal restrictions, and other land information in the Netherlands, Eastern Europe, and numerous developing countries. Currently, his main focus is on innovative land tools, especially to expand tenure security to the legitimate, previously unrecorded rights of the poor and underprivileged. In addition to teaching, supervision (40 MSc and 15 PhD graduates), and research, he also is a member of the International Advisory Board of the Global Land Tool Network (GLTN) and the Advisory Board of LANDac, the Dutch Academy for Land Governance. Based on his dual disciplinary background of an MSc in geodesy (TU Delft, 1990) and an LLM in Dutch Law (Leiden University, 1992), he wrote an interdisciplinary PhD on systems of land registration (TU Delft, 2002) and ever since his student years has published and reviewed in the interdiscipline aspects of land administration.

Walter T. de Vries is an associate professor and chair of land management at the Technical University Munich in Germany. With an MSc in geodesy and a PhD in public administration, his research deals with the implementation and effects of land information infrastructures and capacity development for cadastres and land administration organizations. In addition, he is currently senior editor of the *Electronic Journal of Information Systems in Developing Countries* (EJISDC). He manages a variety of courses in land administration and geo-information management, has supervised over 80 MSc students for the past 20 years, and has more than 25 years of experience in development assistance projects in Africa, Asia, and Latin America. These collaborative projects were typically with ministries of land, cadastres, and national mapping agencies and usually connected to educational development at universities and/or polytechnics teaching land administration and management.

Rohan M. Bennett is an assistant professor at the University of Twente in the Netherlands. He has experience in land administration design and assessment in countries across sub-Saharan Africa and Asia. His research focuses on better aligning the societal demands of tenure security, food security, and climate change with the technological opportunities presented by low-cost global navigation satellite systems (GNSS), unmanned aerial vehicles (UAVs), and cloud-/crowd-based geo-services. He holds degrees in geomatic engineering and information systems from the University of Melbourne, Parkville, Australia, and earned a PhD from the same institution in 2008. He acts as a reviewer for over 20 journal and conference series. Since 2010, he has co-supervised 20 successful MSc theses and 3 PhD theses.

Contributors

Berhanu K. Alemie
Institute of Land Administration
Bahir Dar University
Bahir Dar, Ethiopia

Danilo Antonio
Land and GLTN Unit
Urban Legislation, Land and Governance Branch
United Nations Human Settlement Programme
UN-Habitat/GLTN
Nairobi, Kenya

Clarissa Augustinus
Land and GLTN Unit
Urban Legislation, Land and Governance Branch
United Nations Human Settlement Programme
UN-Habitat/GLTN
Nairobi, Kenya

Rohan M. Bennett
University of Twente
Enschede, The Netherlands

Mireille Biraro
INES-Ruhengeri (Institut d'Enséignement Supérieur de Ruhengeri)
Ruhengeri, Rwanda

Luc Boerboom
University of Twente
Enschede, The Netherlands

Arnold Bregt
Laboratory of Geo-Information Science and Remote Sensing
Wageningen University
Wageningen, The Netherlands

Kees Bronsveld
University of Twente
Enschede, The Netherlands

Walter T. de Vries
Technical University Munich
Munich, Germany

Peter Fosudo
Lade Fosudo & Co.
Lagos, Nigeria

Yola Georgiadou
University of Twente
Enschede, The Netherlands

Liza Groenendijk
University of Twente
Enschede, The Netherlands

Solomon A. Haile
Land and Global Land Tool Network Unit
Urban Legislation, Land and Governance Branch
UN-Habitat
UN-Habitat/GLTN
Nairobi, Kenya

Potel Jossam
Land Office Kyonza District
Government of Rwanda
Eastern Province, Rwanda

Susan Keuber
Mukono District Land Office
Mukono, Uganda

Peter Laarakker
Kadaster International
Apeldoorn, The Netherlands

Christiaan Lemmen
Kadaster International
Apeldoorn, The Netherlands

Monica Lengoiboni
University of Twente
Enschede, The Netherlands

Co Meijer
Kadaster International
Apeldoorn, The Netherlands

Joe Nasr
Independent International Consultant
Toronto, Ontario, Canada

Bert Raidt
University of Twente
Enschede, The Netherlands

Georgina Rockson
Office of Administrator of Stool Lands
Ghana

Libia Y. Romero Lara
São Paulo, Brazil

Didier Milindi Rugema
Rwanda Natural Resources Authority
Kigali, Rwanda

Marie Christine Simbizi
University of Twente
Enschede, The Netherlands

Dimo Todorovski
University of Twente
Enschede, The Netherlands

Paul van der Molen
University of Twente
Enschede, The Netherlands

Jeroen Verplanke
University of Twente
Enschede, The Netherlands

Fikerte A. Yimer
Ethiopian Mapping Agency
Addis Ababa, Ethiopia

Jaap Zevenbergen
University of Twente
Enschede, The Netherlands

Section I

Introduction

1

Toward Responsible Land Administration

Walter T. de Vries, Rohan M. Bennett, and Jaap Zevenbergen

CONTENTS

Introduction

The motivation of the coeditors of this book came from several observations and experiences in land administration throughout the world. The first is the recognition that many titling programs do not seem to generate the anticipated results, either because the speed with which titles or some form of land rights are delivered is far from sufficient or, perhaps more so, because many of the conventional approaches seem to fail in providing a minimum form of tenure security for the most vulnerable (Zevenbergen et al. 2013). Estimates suggest that less than a quarter of the countries in the world maintain complete land administration systems. This would translate into an estimate of 4 billion of the world's 6 billion land tenures remaining outside formal governance arrangements. In these cases, information about people and the land they use remains unrecorded and obscure to governments, firms, or citizens, and reversely, citizens, firms, or governments cannot legally claim their rights to land. The situation prolongs the conditions of the already 1 billion people who live in slums with insecure tenure and of rural migrants entering these informal settlement areas every day, particularly in sub-Saharan Africa.

The second observation is that the current situation impedes all sorts of development activities: land tenure insecurity enables land grabbing and promotes land disputes; land value uncertainty impedes markets and tax governance; land use and development activities (e.g., land readjustment and consolidation) for food security and climate change can neither be designed nor implemented properly. The gap between the land administration systems of more and less developed countries is referred to as the *cadastral divide* (Bennett et al. 2013). Efforts to bridge the divide span well over half a century, however, ultimately have failed: conventional Western approaches cannot be easily transplanted into other parts of the world. Or, putting it more strongly, these approaches should not be transplanted to other countries. Alternatives are needed, which are much more adapted to local needs and circumstances and which are grounded in legitimate and acceptable societal and institutional practices and customs.

Against this background, new insights on what can be considered legitimate land rights and alternative approaches for documenting those have been emerging in land administration science. These focus on generating tenure security for all, tend to build on a combination of faster and cheaper technical options, and acknowledge different forms of land tenure. On the one hand, emerging geospatial technologies, such as low-cost Global Navigation Satellite System (GNSS) tools, high-resolution satellite imagery, and unmanned aerial vehicles (UAVs), are offering new techniques for data collection and management. On the other hand, new insights in acknowledging and securing land rights, such as the continuum of land rights, the recognition of family and customary rights, and the voluntary guidelines on responsible tenure offer alternatives for legal records.

Equally, new societal drivers including food security, rapid urbanization, climate change, and post-conflict redevelopment are driving the need for innovation in social approaches and forms of inclusion in land administration systems. Such approaches take the specific needs of land users and land tenants better into account. These also align technical and administrative requirements of land administration better to the social and legal requirements. The democratization of information and communication technology, as well as a growing number of community and individual initiatives to record land rights in different manners, require new types of solutions and new ways to develop capacity. Collectively these new opportunities set out a shift in land administration design and management toward, what we call in this book, *responsible* land administration.

Theoretical Notion of Responsible Land Administration

Responsible land administration is a construct that challenges conventional forms of land administration. There is a need to establish such a new notion, because conventional land administration paradigms are often rooted in historically developed views on land management. Hereby, securing tenure has often been equated with organizing the management of transactions with land and real estate. In this perspective, guaranteeing mostly individual land rights can be handled most efficiently if it is a statutory responsibility of one or more state-based organizations that are entrusted to handle large-volume registers and databases and/or manage associated bylaws and operational procedures. Though such systems have long been looked upon as the result of promising disciplinary innovations and improvements, they have failed at times to be effective and sufficiently authoritative because they ignored ethical and societal questions until it was too late. In contrast, responsible administrative systems are those that incorporate innovations with an understanding of the possible ethical and societal implications.

The element of responsible adds a new notion to conventional approaches in land administration. It aligns the administration much more fundamentally to the ever-changing needs and capabilities of individuals, government, and society. Bourgon (2007) considers this innovation even a real paradigm shift in public administration, and also considers this part of a trajectory of governments and their administrative agencies and individuals becoming more *responsible, responsive,* and *respected.* Responsible in this context means the need to be involved in building collaborative relationships with citizens and groups of citizens, encouraging shared responsibilities, disseminating information to elevate public discourse and foster a shared understanding of public issues, and seeking opportunities to involve citizens in government activities (Bourgon 2007). Cooper (2012) further adds that a responsible administrator should not only be driven by external and internal organizational obligations (usually the choice among available resources and handling resource constraints), but also—and perhaps more so—by the "tension between public and private interests and the quandaries over how to manage them" (Cooper 2012). In other words, administration is only responsible when it continuously aligns internal processes and resources with the dynamics of societal demands.

The term "responsible" was introduced as well in the voluntary guidelines on the responsible governance of tenure of land, fisheries, and forest in the context of food security (FAO 2012). Responsible in these guidelines is a specific attribute that encompasses socioeconomic development, poverty eradication, and food insecurity. It is a type of governance that recognizes inherent human dignity and equal and inalienable rights of all individuals; adheres to nondiscrimination; stimulates equality between individuals by acknowledging differences between individuals and taking positive action to promote equitable rights, ensuring equal rights of women and men; prioritizes holistic and sustainable approaches; works on the basis of consultation and participation; adopts approaches that are transparent, rule based, and applicable for all; and aims for continuous improvement strategies by monitoring, analysis, and evidence.

Taking into account these shifts toward responsible government and governance, the essence of the adjective responsible for land administration is therefore the adoption of a collection of practices in relation to land. These include the following:

1. Multi-stakeholder focus: Acknowledges and valorizes the representation of relevant stakeholders in the development and/or innovation of land policies and land administration systems. These may include both formal land right holders and nonregistered, yet legitimate, land tenure holders. Stakeholders need to be represented and involved in consultation and valorization panels such that results of innovation can be implemented collaboratively.
2. Multidisciplinary: Uses and adopts insights from multiple disciplines and, in particular, connects innovations from technical and informational sciences (e.g., land surveying/geodesy, computer science, informatics, information sciences, agriculture) to the social sciences and humanities (law, anthropology, development studies). Innovations in land administration tend to be cross-disciplinary, that is, using insights of multiple disciplines simultaneously, or even transdisciplinary, that is, transforming insights of a single discipline by the new insights from another discipline.
3. Proactive: Ethical and societal aspects are incorporated into the design process from the start.
4. International: Research has an international focus and takes the global context into consideration.
5. Relevant and usable: Research development will be considered in terms of societal relevance and practicality of implementing the knowledge.

Aspects of Responsible Land Administration

So, how do these characteristics of responsible governance and innovation make sense in the day-to-day reality of land administration and land administrators? This book approaches this question from three angles:

1. New drivers and inspirations changing land administration: This includes a specific bias in providing tenure security and designing land administration system toward the pro-poor, connecting land administration to specific challenges of providing food security, constructing land administration systems that align with dynamic processes of urbanization and allow for effective and responsible

urban governance, reconstructing and establishing land administration systems during and after major conflicts, and assembling land administration systems from the crowd in the cloud.

2. Innovative technical and operational designs of land information systems: These are not only based on opportunities arising from new technologies, but also fitting particular societal challenges. Such new solutions for land information systems include the point cadastre, the digital pen toolkit for participatory mapping and quick and accurate cadastral data acquisition, the speed updating of cadastral records, the fit-for-purpose land consolidation, and the dynamic/nomadic cadastre.
3. Impacts of land administration systems and the new ways to measure these impacts: Evaluative studies reveal how, where, and why formalizing land tenure affects and triggers land use changes; similarly, how, where, and why land administration systems interlink and interact with systems of environmental protection; how, where, and why land displacement aligns or disarranges land administration; and what are the interrelations between systems of land leases and urban development/urban land management.

All chapters are grouped under one of these three angles. Each of the chapters is built on a theoretical grounding and empirical research, and each is structured so that not only are the new challenges, design, or impacts presented, but also the theoretical or scientific context are explained and justified. In this way, the practical experience and the empirical observations are systematically connected and extended to new scientific insights. Though we believe that the insights provided in this book are relevant globally, we have opted in this book to zoom into cases of sub-Saharan Africa, and in particular to central and eastern Africa where most of the innovations in land administration are currently implemented and tested.

Methodological Diversity and Rigor

As land administration is a scientific field that operates at the cross points of other disciplines, notably land surveying, geodesy, (land) law, planning, public administration, (land) politics, land economics, development studies, and anthropology, there is no single research framework that has proven to be useful to study all aspects of land. Though this seems a challenge for a researcher in land administration, this book aims to overcome the gaps between the disciplines and associated approaches and contribute to the still underdeveloped land administration research in and for countries in transition by

1. Balancing the Western-/Northern-centric focus of land administration systems and paradigms with more attention to the innovative land administration systems being developed in countries in transition in the global south, and in particular in Africa.
2. Incorporating studies at subnational scale and recognizing the substantive relevance of regional and local levels.
3. Linking specific administrative circumstances and technological design on the one hand, and performance (efficiency, administrative transparency, sustainability, or any other criterion) on the other hand. This requires the fusion of theoretical perspectives from both social and technical sciences.
4. Emphasizing theory-based and empirically grounded research with rigorous methodological approaches that more adequately reflect the interdisciplinary nature of land administration and the interplay between social (humans, organizations, and institutions) and technical issues.

The aim of the book is to redress the limitations of conventional land administration research, advance the scientific discourse, and contribute to practice. This book sets the scene to encourage theoretically and empirically sound land administration research that can contribute to better informed and more useful developments in responsible land administration. This should explain why and how land administration is taking different forms and why they succeed or fail in various circumstances around the globe. In-depth research that builds on a variety of disciplinary knowledge relates to a theoretical framework, and includes empirical references that transcend descriptive and normative insights, and attempts to go beyond the *what* and *who* and answer the *how* and *why*. Finally, by promoting the comparative approaches that point to key system-related determinants of land administration, the book aims to gain better insights into those fundamental matters.

Structure and Content of the Book

Along with this introductory section and the future orientated conclusion, the book contains three major sections, each with several chapters. The sections follow the three main aspects of responsible land administration:

1. New inspirations
2. Creating innovative designs
3. Measuring the impacts

A fourth section complements these by looking ahead.

New Inspirations

This section combines five fundamental drivers and challenges. Simbizi, Bennett, and Zevenbergen analyze in Chapter 2 how land administration systems could better respond to the needs of society's poorest. This chapter provides a background to the discourse and practical developments: the convergence of pro-poor theories, land tenure security, and fit-for-purpose land administration is explained, as is the continuum of land rights. Three analytical frameworks are used to illustrate how pro-poor land administration is applied in practice. Empirical data from recent fieldwork in Rwanda are used for the analysis. Pro-poor land administration is shown to be implementable at national scales; however, implementation of pro-poor design elements does not guarantee improved land tenure security for all the poor: other socio-technical elements are crucial to consider including perceptions, equality, multiple relationships between individuals, social institutions, public institutions, the continuum of land rights, and land information systems.

Bennett, Rockson, Haile, Nasr, and Groenendijk address in Chapter 3 the interactions between land administration systems and food security programs. Good systems of land administration are often argued as an important enabler of food security: land information systems support small-scale access to credit, investment in agriculture, make possible land consolidation activities, and protect the vulnerable in large-scale investment activities. However, land administration can equally play a negative role if not implemented responsibly. A model is developed that aims to better articulate the multiple links between land administration and food security. It can be used as a starting point for cohesive development of multipronged intervention strategies that ensure positive links between people, land, and their food.

Alemie, Bennett, and Zevenbergen connect in Chapter 4 land administration to urban governance. The chapter brings together two disparate discourses and takes a balanced view on the role of land administration in delivering *good enough* urban governance. The rapidly urbanizing cities of Ethiopia are focused on to provide context. A conceptual framework linking policy agendas, legal frameworks, urban land administration systems, and multiple stakeholders is developed. Urban land administration systems are argued as a theoretically neutral concept: they can play either a supportive or an undermining role in urban land governance. The conceptual framework, when applied in practice, will assist in assessing challenges and opportunities relating to existing or planned urban land administration systems.

Todorovski, Zevenbergen, and Van der Molen provide in Chapter 5 evidence of how land administration can play a pivotal role in post-conflict state building. This relation is explored empirically through evaluating the post-conflict state-building process in Rwanda. This chapter starts from the premise that core land administration elements can be built into broader state-building processes: land sharing processes, villagization projects, and allocation of state land for returnees and that land registration programs can be interrelated with other programs. The chapter addresses whether land administration can be recognized as a facilitator of post-conflict state building. For the case of Rwanda, part of the answer lies in the historical background of the violent conflict and in further understanding the original displacement and the challenges faced by returnees. Overcoming both types of predicaments adds to the overall process of post-conflict state-building and land administration development in Rwanda.

Laarakker, Zevenbergen, and Georgiadou conclude the section on inspirations in Chapter 6 on crowds and clouds in land administration. This chapter argues the next wave of technological innovation is being sparked by *the cloud* and *crowdsourcing*. The cloud enables vulnerable land records to be stored and backed up in multiple locations at low cost: the long-standing issue of keeping land information systems maintained in less developed locations now has a more transparent and low-cost solution. Meanwhile, crowdsourcing enabled by smartphones is now prevalent across much of the developing world. It enables non-surveyors to map and record their own land parcels: the decade's old problem of expediting land rights recording may soon receive a boost. However, the innovation comes with a double edge: issues of information security, privacy, accuracy, and reliability must also be resolved. This chapter looks at developments in crowd- and cloud-based land administration, how they are changing land administration thinking and design, and how the potential drawbacks might be resolved.

Creating Innovative Designs

In Chapter 7, de Vries, Meijer, Keuber, and Raidt address point cadastres. A point cadastre is a stripped-down method for collecting and maintaining cadastral data. Geographic points, not parcels, are used as the key reference to represent land parcels in a cadastral database. The approach is suitable where weak interests in land prevail, informal tenure is not aligned with a formal registration system, or connections to other registers are lacking. This chapter deals with how to build a point cadastre and identifies the challenges needing to be overcome. Empirical data are drawn from recent work on Bugala Island in Uganda and Bissau, the capital of Guinea-Bissau.

Rugema, Verplanke, and Lemmen address the potential of digital pens for cadastral surveys in Chapter 8. Using tests in Rwanda, this chapter introduces an alternative to populating a land administration system: the digital pen toolkit can reduce the workflows for capturing parcel geometry. It is a low-cost technology that can be operated with relatively limited training. Instead of acquiring data through a participatory mapping process using ortho-photos, regular drawing equipment, and extensive post-processing, the use of a digital pen allows storage of the digital data directly into geo-referenced digital format. When comparing this to the analog method it reduces sources of errors, saves time, reduces archiving space, and optimally utilizes the benefits of participatory mapping.

In Chapter 9, Biraro, Bennett, and Lemmen address the problem of updating land administration systems. This chapter explains the contemporary emphasis on updating: systems should be cheap and easy to maintain. Rwanda completed registering all its lands through a systematic land registration program using the general boundary approach. The work was completed in a period of 5 years. However, to be sustainable, the program still required considerable effort and well-designed procedures for updating land information. The case of Rwanda was used to examine the obstacles and technological alternatives for keeping land information up-to-date. These include parallel use of locally based one-stop shops, low-cost eGovernment alternatives, and integrated land information infrastructures.

Bennett, Yimer, and Lemmen describe fit-for-purpose land consolidation. Land consolidation is the process of reallotting or redesigning rural land holdings to enable more efficient agricultural production and improved social conditions. Up until recently, land consolidation remained an activity practiced in western European countries: the potential for its application in less developed contexts, supported by land administration processes and data, is only now being explored in sub-Saharan Africa. Using a socio-technical approach, this chapter investigates the land consolidation innovations in European countries and explores how they might be tailored to suit countries elsewhere. Processes from the Netherlands and the application case of the Amhara region in Ethiopia are focused on. The resulting showcase demonstrates how fit-for-purpose land consolidation can be achieved when digital cadastral data and simple GIS tools are available, however, even simpler tools that enable more participation from local smallholders would be preferable.

The section of innovative designs is concluded by Chapter 11 from Lengoiboni, Bregt, and van der Molen containing a study on dynamic/nomadic cadastres. Conventional notions of the *land parcel* have been extended: previously unrecognized tenures including customary, nomadic, or communal interests are now incorporated into the concept. Technical tools including the Social Tenure Domain Model (STDM) enable these new understandings to be operationalized in land administration systems. The nomadic pastoralists of Kenya's dryland regions illustrate where these new approaches can be applied. Example, Kenyan counties are discussed to illustrate the issues and confirm where such approaches have utility. It is shown that the interaction between conventional and pastoralist tenures needs more assessment from spatial, legal, economic, environmental, and social tenure perspectives. These can be used for enabling innovative recording of pastoralist tenures, and improving awareness between stakeholder groups.

Measuring the Impacts

Fosudo, Bennett, and Zevenbergen open this section with Chapter 12 on land administration effects on land-use change. In some cases, securing land tenure through regularization is assumed to sustain rather than to change land use. However, this assumption may not hold in areas of considerable land-use dynamics, such as urban fringes. The peri-urban areas of Kigali city in Rwanda are exemplary for such contemporary dynamics, in particular in relation to agricultural land use changing to residential land use. At the same time Rwanda has strongly reinforced a land tenure regularization (LTR) program fostering more secure land tenure. This chapter shows how detecting the change in land use is possible using a combination of remote sensing, cadastral data, and ground proofing activities. The empirical comparison was done for the period 2008–2013. Over the study period, a loss of approximately one-third of agricultural parcels occurs over 103 ha. These dramatic reductions in agricultural land use, simultaneous with the LTR program, suggest a significant impact of land administration programs on land use; however, other factors are also at play.

In Chapter 13 Romero, Zevenbergen, and Bronsveld debate how environmental protection can be achieved via land administration. The establishment of protected areas of land preserves biodiversity and wildlife, yet tends to neglect rights of local communities, especially in national parks. This chapter examines the challenge for land administration systems to deal equitably with the potentially competing rights of local communities and the environment. Tanzania has broadly embraced protection policies, yet has also seen conflicts of interests. In Saadani Park, Tanzania, one of the most recently gazetted national parks, local communities claim diverse rights in the neighboring area, such as the villages Saadani, Uvinje, and Bujuni. The analysis reveals that stakeholders only diverge in the actions that should be taken, not in the goals. One of the conflicting strategies was "keeping things as they are," which would lead to unfair compensation, loss of rights to land, and community displacement. With this knowledge, potentially captured in a land administration system, it is possible to derive alternative strategies and instruments to reduce the conflicts between people, environment, and land rights.

Jossam, van der Molen, Boerboom, Todorovski, and de Vries address in Chapter 14 the land displacement vis-à-vis land administration. This chapter conceptualizes displacement and post-conflict land administration, and evaluates how land administration is handled in emergency, early recovery, and reconstruction periods of conflict situations. Evidence from post-conflict Rwanda, whose government envisaged good land governance as a contributor of sustainable peace and security, informs the analysis. A nationwide systematic land registration program aimed to register land all over the country. The program faced many challenges including continuous competing land claims and disputes. These long-standing problems are a result of the way in which land issues were handled in the emergency and early recovery period, especially during the land sharing period and imidugudu (collective settlement policy).

Chapter 15 by Antonio, Zevenbergen, and Augustinus rounds off the section by an evaluation of the STDM software as piloted in Uganda's Mbale municipality. STDM is an open-source land information management tool that can be used by poor communities and their networks with little support from land professionals to support localized land management. STDM can support bridging the cadastral divide, from the bottom-up. The case reveals the opportunities and challenges for STDM implementation in Uganda and beyond; mobilization, sensitization, customization, and stakeholder engagement are key issues. A greater challenge is to convince existing land institutions of the utility and potential of the tool, not only as one that can be used for land information management, but also as a tool that facilitates improved land governance more generally.

Looking Ahead

The concluding chapter by Zevenbergen, de Vries, and Bennett looks ahead based on all contributions of this book. The chapter synthesizes the key concepts, ideas, and findings from the book. The shift in land administration, visible in the new societal drivers, the emergence of new technological designs and innovative methods for evaluation, sets the scene for a new kind of land administration: responsible land administration. Responsible land administration is especially developing in the context of the Anthropocene—an urban age where global challenges of rapid and massive scale urbanization and migration coincide with major conflicts relating to land, food security, water, infrastructure, and other resources. The future of land administration has to address five types of changes that are currently emerging or gradually occurring: in people-to-land relations, in conceptual understanding and technological possibilities, in land use, in measuring methodologies, and in (change) agents. Land administration needs to scale-up and integrate with other domains, and incorporate new axioms, paradigms, and research questions. Such research questions are mainly socio-technical and institutional in nature, creatively combining globally available technologies and technical platforms with a clear understanding of a legal, organizational, and management context. Teaching, or better, capacity development, for the reinvented *responsible land administrator* calls for not only understanding the interdisciplinary nature of land administration, but also of becoming change agents that design, convince politicians and interdisciplinary staff of its use and get it implemented and evaluated. In this way, land administration can develop into a new type of scientific discipline that can support the derivation of contemporary fit-for-purpose and responsible solutions.

References

Bennett, R.M., H.A.M.J. van Gils, J.A. Zevenbergen, C.H.J. Lemmen, and J. Wallace. 2013. Continuing to bridge the cadastral divide. *Proceedings of Annual World Bank Conference on Land and Poverty*. Washington, DC: World Bank.

Bourgon, J. 2007. Responsive, responsible and respected government: Towards a new public administration theory. *International Review of Administrative Sciences* 73(1): 7–26.

Cooper, T.L. 2012. *The Responsible Administrator: An Approach to Ethics for the Administrative Role*, 6th edition. First original in 1982. Hoboken, NJ: Wiley.

FAO. 2012. *Voluntary Guidelines on the Responsible Governance of Tenure of Land, Fisheries and Forests in the Context of National Food Security*. Rome, Italy: FAO.

Zevenbergen, J., C. Augustinus, D. Antonio, and R. Bennett. 2013. Pro-poor land administration: Principles for recording the land rights of the underrepresented. *Land Use Policy* 31: 595–604. doi: 10.1016/j.landusepol.2012.09.005.

Section II

The New Inspirations

2

Pro-Poor Land Administration

Marie Christine Simbizi, Rohan M. Bennett, and Jaap Zevenbergen

CONTENTS

Introduction

Conventional land administration systems often work against the needs of the poor (Lemmen 2010; Zevenbergen et al. 2013). In sub-Saharan Africa the approach has failed the poor, even for those projects considered partially successful (Van Asperen and Mulolwa 2006; Zevenbergen et al. 2013). Concerns abound that conventional survey and mapping requirements are expensive, timely, and do not align with existing contextual capacity (Enemark et al. 2014). This results in popular estimates that 75% of the global population does not have access to a formally recorded and recognized land right: a land administration divide is evident (Bennett et al. 2013). This is important: property is the basic legal concept on which conventional legal systems are based and on which subsequent human rights are derived, regardless of whether the property is held by an individual, a community, a family, tribe, or clan (Van Der Molen 2006).

In response to this land administration divide, a new era of so-called pro-poor approaches has emerged. These aim to ensure land administration systems responsibly deliver land tenure security to the poor. The term "pro-poor" is not new: it defines approaches that take into account people living in poverty and was first used in the context of slum dwellers (UN-Habitat 2007, 2008; Van Der Molen et al. 2008). The approach was extended to the rural poor in parallel. Though there is no agreed definition of pro-poor land administration, there appears to be consensus in the underlying principles and strategies. For instance, most discussions tend to mention the continuum of land rights, gender equality in relation to land access and tenure security, affordability and low-cost land registration, local knowledge as an important resource for land registration, localized approaches and specific tools, and flexible approaches to land management, among others (UN-Habitat 2007; Payne et al. 2009; Williamson et al. 2010). These principles need to be underpinned by pro-poor tools and approaches to support pro-poor land management (Williamson et al. 2010). Consequently, initiatives such as the Global Land Tool Network (GLTN) were established with a specific mandate to develop pro-poor land management tools. A range of other initiatives from industry, donors, and academia are also evident. These include conceptual tools related land tenure, land policy tools (Van Der Molen and Lemmen 2006; UN-Habitat 2007, 2008; Enemark et al. 2014; Simbizi et al. 2014; Whittal 2014), and more technical tools intended to support land tenure mapping and land registration (Lemmen et al. 2009; Zevenbergen et al. 2013).

Application of the pro-poor land administration approach is already evident. Particularly over the last decade, governments of many sub-Saharan countries commenced reengineering existing land administration systems to include a pro-poor mindset. Institutions involved in land governance have been redesigned, land policies and laws redrafted, and surveying and mapping techniques radically simplified to respond to the pro-poor mindset. Recent efforts include Rwanda's Land Tenure Regularization (LTR) program (Sagashya and English 2010), the Rural Land Certification programs in Ethiopia (Deininger et al. 2008), the Land Administration Reform in Ghana (Independent Evaluation Group [IEG] 2013), the Land Tenure Services Project in Mozambique (Hagos 2012), and land reform in South Africa (Benjaminsen et al. 2009). Despite all these examples, whether the pro-poor mindset actually translates into improved tenure security for the poor is still unclear.

In this regard, this chapter aims to provide insights on how pro-poor land administration works in practice, and whether prescribed principles, when followed, actually deliver improved land tenure security for the rural poor. To achieve this, the chapter makes specific use of the recent Rwandan LTR program. In 2007, LTR was launched via pilots and covered the entire nation by the end of 2013. The program was implemented with the desire to provide land tenure security to all Rwandans (Sagashya and English 2010). LTR is perhaps the preeminent showcase of an intervention designed with the pro-poor land administration mindset. The intervention is considered to have accommodated most pro-poor ingredients, as prescribed in the contemporary literature (Lemmen 2010; Zevenbergen et al. 2013; Enemark et al. 2014): a general boundary approach was used, ortho-photos were used for boundary demarcation and recordation into the land information system, a flexible land law that recognized existing land rights was used, and gender issues were addressed within those laws. This study focuses specifically on the rural poor who constitute the majority of the Rwandan population, 83.5% according to recent population and housing census (Government of Rwanda 2013), but, also 70% of the world's poor population are rural dwellers. Furthermore, it was estimated that at least 70% of the world's very poor people are rural and the large proportion of these are in South Asia and sub-Saharan Africa. The rest of this chapter is structured as follows. A theoretical background to pro-poor land administration is provided. This leads to an overview of the methodology, an evaluation that utilizes three pro-poor land-related frameworks. The results of the evaluation are then presented and discussed. Conclusions look at the implications of the results for future practical and research endeavors.

Theoretical Perspective

Until recently, land administration systems in many sub-Saharan African countries were based on what Williamson and Ting (2001) describe as a narrow land administration paradigm; one introduced during the colonial era centered upon conventional land registration and cadastral mapping techniques. The approach was not practical: land titling tools developed in Western Europe and were not conducive to local conditions and capacities (Williamson et al. 2010; Zevenbergen et al. 2013). Despite the inadequacies, the approaches persisted into the contemporary era: vested interests seek to maintain a status quo that often favors the elite within the country. Consequently, many land polices, laws, and procedures are still biased against the poor (UN-Habitat 2007). It is wrong to assume the governments undertaking land formalization activities are easily able to establish pro-poor institutions and new forms of formalized rights (Sjaastad and Cousins 2009).

The pro-poor land administration movement emerged from various sources in response to inadequacies in land tenure security. Pressure came top-down from the international land sector and bottom-up from local NGOs and farming groups; more effective forms of land administration that serviced the poorest in a community were needed (Van Der Molen and Lemmen 2006; Enemark et al. 2014). These approaches were seen as the way forward to providing land tenure security to the poor.

Contemporary literature describes what a pro-poor land administration system entails. Features or ingredients can be grouped into two categories. The first involves technical tools for spatial data acquisition and recordation. It is recommended that affordable technology is used to build the underpinning land administration spatial framework (Enemark et al. 2014). Tools such as aerial photographs and satellites images are considered to be cost-effective for large-scale spatial data acquisition (FIG 1999; Williamson and Ting 2001; Tuladhar 2005; Zevenbergen et al. 2013; Enemark et al. 2014). In addition, a land information system should be able to accommodate a variety of spatial units and land tenure arrangements (Van Der Molen and Lemmen 2006; Zevenbergen et al. 2013; Enemark et al. 2014). The second category concerns pro-poor legal and policy tools. Several researchers and international organizations argue that national land policies and existing legal frameworks should recognize and promote the recordation of customary land rights (FAO 2002; Deininger 2003; Van Asperen and Mulolwa 2006; Van Der Molen and Lemmen 2006). It is suggested that a pro-poor land policy can provide a range of land rights tailored for different situations (UN-Habitat 2007). The same land policy should insure that the poor have access to land and that land services are set at prices that can be afforded. Borras and Franco (2008) disqualify land policies that give superficial formal land rights, but not the power to exercise control and management of use. They stress that pro-poor land policies should involve real material gain to be retained or transferred to the poor. Overall, the consensus is that the ultimate pro-poor solution is to shift toward unconventional land administration approaches (Van Der Molen 2006; Van Der Molen et al. 2008).

Along with the pro-poor movement, theoretical research has revisited the notion of land tenure security to insure that it is sensitive to the needs of the poor (Lavigne-Delville 2006; Van Gelder 2010; Arnot et al. 2011; Simbizi et al. 2014). In general, it is argued that the conventional meaning of land tenure security (i.e., individual land ownership, boundaries accurately surveyed, proof of land ownership, degree of exclusivity, and duration of land rights) (Feder and Feeny 1991; Platteau 1996) does not align with the land tenure systems of the poor in developing countries (FAO 2002; Deininger 2003; Lavigne-Delville 2006; Simbizi et al. 2014).

A synthesis of the abovementioned literature reveals the conceptual framework on which this chapter is based (Figure 2.1). The ingredients of pro-poor land administration, as derived from the fit-for-purpose approach (Enemark et al. 2014) and the design elements of a pro-poor land recordation system (Zevenbergen et al. 2013), act as a tangible basis. In addition, the components of land tenure security for the rural poor in sub-Saharan Africa are also incorporated (Simbizi et al. 2014).

The choice of the three conceptual frameworks combined in Figure 2.1 is motivated by several reasons. First, the three theoretical works constitute well-recognized components of the pro-poor land administration domain: (1) fit-for-purpose land administration considers the entire land administration system, (2) the pro-poor land recordation system elements focus specifically on land recordation activities, and (3) the model of rural poor land tenure security looks at the actors and entities involved in the process. Second, the three frameworks are able to be integrated: the different aspects that each deals with, previously studied in isolation, can be drawn together in this new conceptual model. Third, the fit-for-purpose approach (a result of collaboration between the World Bank and the International Federation of Surveyors [FIG]) and the pro-poor land recordation system (backed by the GLTN and the UN-Habitat) are products of arguably the most active and influential institutions in the field of land administration. This explains why the two approaches have received an increasing acceptance and support of donors and governments. Although available research results provide

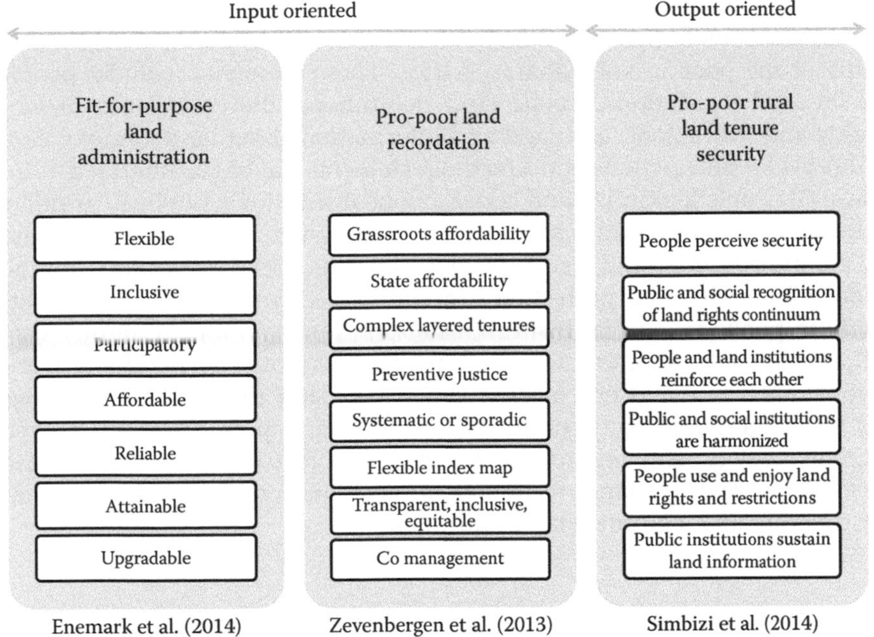

FIGURE 2.1
Frameworks for pro-poor land administration and land tenure security provision. (Adapted from Enemark, S. et al., *Fit-for-Purpose Land Administration: Open Access e-book*. International Federation of Surveyors (FIG), Copenhagen, Denmark, 2014; Zevenbergen, J. et al., *Land Use Policy*, 31, 595–604, 2013; and Simbizi, M.C.D. et al., *Land Use Policy*, 36, 231–238, 2013.)

successful stories on the implementation of the two approaches, few attempts are solely dedicated to examining the effects of the two approaches against the situation of tenure security of the poor.

The fit-for-purpose and the pro-poor recordation system prescribe *inputs* to land administration systems that are believed to improve the land tenure security of the poor. For example, the demand for flexible and affordable approaches to spatial information recording, and a flexible legal framework that accommodates society, can be considered in this light. The pro-poor land recordation system is more specific on this point: the system should be affordable and accessible to a society's poorest members. Similarly, the state should be able to afford the cost to setup and sustain the system in reasonable amounts of time. That is, at least from the fit-for-purpose perspective, the system should be *attainable* and *reliable*. Both fit-for-purpose land administration and the pro-poor land recordation system promote *inclusive* and *participatory* approaches to recording spatial data on land and associated legal and social tenure rights. These include, but are not limited to, the continuum of land rights approach. The later implies recognition of different forms of land tenure, other than individual freehold.

The content of land tenure security presented in Figure 2.1 is derived from the interactions between five main elements found in the land tenure systems of the poor in sub-Saharan Africa. Those elements include: people, social land institutions, public land institutions, the continuum of land rights and restrictions, and land and information about land (Simbizi et al. 2014). Six key interactions between those elements can be summarized as follows: (1) people legitimize and access public institutions: public institutions allocate land with equity, and protect and recognize people and their land; (2) people legitimize and access social institutions: social institutions allocate land with equity, and protect and recognize people and their land rights; (3) both public and social institutions are harmonized and recognize each other; (4) people hold, use and enjoy their land rights and restrictions, and people are aware and empowered about their land rights and restrictions; (5) land rights and restrictions are balanced, and public institutions recognize, protect, and enforce land rights and restrictions; and (6) social institutions regulate land; recognize, protect, and enforce land rights and restrictions; and legitimize land information.

Methodology

In line with the aim of the chapter, the Rwandan LTR is used as a basis for evaluating how pro-poor land administration approaches are implemented in practice, and the extent to which they deliver land tenure security to the poor. Three frameworks were used in the evaluation: (1) the fit-for-purpose land administration characteristics of Enemark et al. (2014), (2) the pro-poor land recordation requirements of Zevenbergen et al. (2013), and (3) framework of rural poor land tenure security of Simbizi et al. (2014). First, using Simbizi's framework, the land tenure security of the rural poor subsequent to LTR implementation is assessed. Empirical data collected during a related study in 2015 were utilized (Simbizi et al. 2015). Second, the same empirical data, combined with data extracted from a desktop study and government archives, were used to evaluate the extent to which LTR adhered to the ingredients of the frameworks of Enemark et al. (2014) and Zevenbergen et al. (2013) presented in Figure 2.1. The results stemming from each framework are compared, and a more holistic view of the pro-poor approach becomes apparent.

Evaluation activities rely heavily on the techniques, methods, and skills of the underpinning research methodology (Kumar 2005). The case study approach (Yin 2003) was found to be an appropriate method for the task. Case studies (Yin 2003) are known to be a methodology of choice when a holistic investigation is needed. Moreover, case study data collection offers the opportunity to use many different sources of evidence: it enables the use of multiple methods of data collection. For LTR, the main sources of evidence included household surveys conducted in one area that was object of an LTR field trial in 2007, online documentation, archive analysis, and semi-structured interviews.

Data collection specifics were as follows. In 2013, a household survey was conducted in Kabushinge Cell, located in Rwaza Sector, Musanze District, Northern Province. Kabushinge cell was selected as representative of a smallholder land tenure system, and aligned therefore with the focus of this study. The survey was made by means of a questionnaire administered face-to-face. A random sample of 416 households was drawn over a total of 1137 households considered to be the poor. Data entry was made using the software package IBM SPSS 22. Using the same software, statistical analyses including descriptive statistics, cross tabulation, and logistic regression model were performed. Documentation and archive analysis made use of written reports, available manuals, and land-related legal and policy documents. These were reviewed to understand the LTR process, its early effects, and the whole underlying institutional framework. Semi-structured interviews with experts and practitioners were conducted to supplement information obtained from the reviewed documents. Overall, the three sources of evidences covered four units of analysis: plot, household, community, and land administration jurisdiction.

Results

This section presents the evaluation results of LTR against the three pro-poor conceptual frameworks: (1) the content of land tenure security for rural poor (Simbizi et al. 2014) is presented textually first, (2) the fit-for-purpose approach (Enemark et al. 2014) is presented in Table 2.1, and (3) the requirements of the pro-poor recordation system (Zevenbergen et al. 2013) are presented in Table 2.2. Following this, discussions then focus on determining whether the three approaches were adhered to, and if so, whether or not the models support delivery of enhanced land tenure security.

Regarding the status of the *land tenure security of the rural poor* after LTR, the six variables, extracted from *content of tenure security* of Simbizi et al. (2014) and also outlined in Figure 2.1, guide the discussion.

First, regarding people's perceptions on their tenure security, 99% of the study population claim to be satisfied with land administration service delivery. Institution wise, these are good indicators of land institution legitimacy. Indeed, only 15% of respondents fear corruption in public land administration institutions. Against global norms, as found in Transparency International surveys, this is a low percentage. Meanwhile, the cost involved when undertaking land transactions and land conflict resolution processes is perceived

TABLE 2.1

Assessment of Land Tenure Regularization (LTR) through the Lens of Fit-for-Purpose Framework

System Design Elements	Assessment Results: Adherence	Assessment Results: Nonadherence
Flexible	General boundaries utilized for surveying and demarcation. Field identification, adjudication, and demarcation of boundaries utilized rectified aerial photographs and high-resolution satellite images. Both commercial and open-source solutions for spatial data acquisition and processing were used. The maintenance phase utilizes handheld global positioning system (GPS). The new land policy and legal framework provided legal recognition of land rights acquired under customary law.	Land offices at centralized and decentralized level are only equipped with commercial geographic information system (GIS) software (ArcGIS software).
Inclusive	Multiple forms of ownership were included: individual, joint for legally married couples, and group land ownership. All informally held documents (regardless of legal status) were used as evidence during the claim process.	The existing continuum of land rights (including customary rights) was replaced by statutory land ownership rights. The spatial component of the land information system is mainly parcel based.
Participatory	Boundary demarcation was performed by locally trained para-surveyors who were recruited from the community. It was reported that of the 110,000 people who were employed, over 99% were drawn from the community where the work was taking place. Claims over land were assessed in the field.	
Affordable	Implementation cost: land parcel demarcation and adjudication of 10.3 million parcels was achieved at a cost estimated between US$5.47 and US$6.48 per parcel (Gillingham and Buckle, 2014). The implementation cost, though reasonable compared to other countries such as Lesotho (US$69/lease), remains higher than the US$1 per parcel that was achieved in Ethiopia (Deininger et al., 2008).	Without foreign aid and the government support, the country could not have afforded the implementation cost of LTR. Department for International Development (DFID; United Kingdom) was the principle donor contributing more than half of the total cost.

(Continued)

TABLE 2.1 (*Continued*)

Assessment of Land Tenure Regularization (LTR) through the Lens of Fit-for-Purpose Framework

System Design Elements	Assessment Results: Adherence	Assessment Results: Nonadherence
	The land registration fee was set to RWF1000 (almost US$1.5) with a mean plot holding estimated at five per household, the average registration fee is estimated at RWF5000 (~US$7.6).	
	The failure to pay the required fee for land registration has prevented landowners from collecting their land certificates.	
Reliable	Out of LTR process a comprehensive land information system was created.	Keeping the created land information system up-to-date is a clear challenge especially in rural areas. While land transactions taking place in urban areas are reasonably reported, this is not the case in rural areas.
Attainable	The implementation of the system was fast and only required a period of 5 years to achieve nationwide coverage.	
	An estimated 8.4 million leases and freehold titles were prepared with over 5.7 million collected by landowners.	
Upgradable	The established land information system has room for further upgrades. The initial system operates at centralized level, in the future the system could be upgraded to operate digitally, and at a more local level.	

Source: Data collected by authors 2012–2013.

to be too high. The study population was also found to be exposed to the risk of public expropriation and land redistribution. Surprisingly, a small proportion (only 12%) believes in fair compensation. Regarding land institutions (both social and public in nature) allocating land with equity, and recognizing and protecting people and the continuum of land rights, only 52% have land certificates for their total plot holdings. In this regard, LTR appears to have had a double-sided effect on land conflict. On one hand, the process of adjudication has contributed to the boom of land conflict cases in formal

TABLE 2.2

Assessment of LTR through the Lens of Pro-Poor Land Recordation System Requirements

System Design Elements	Assessment Results: Adherence	Assessment Results: Nonadherence
Grassroots affordability	Initial land registration fee (approximately US$1.5 per plot) applied by Land Tenure Regularization (LTR) is not that higher than US$1 suggested for this requirement.	It was found that the poor could not afford the initial land registration fee. The failure to pay the registration fee is one of the reasons that prevented landowners from collecting their land certificate (by 2013 an estimate of 2.7 million of lease out of 8.4 million were not yet collected). Even in the pilot area where the registration fee was waived, 5% of the sample declared having failed to pay the required fee for the plots located outside the pilot area. A land transaction fee (land sale for instance) is perceived to be high (by 81% of the sample). Indeed, the mutation fee in case of a land sale is set to RWF 20,000 (~US$30.76) excluding other associated cost such as transport.
State affordability	The government alone would not have afforded the full cost of LTR (see Table 2.1). Though LTR was implemented using low-cost technology and tools, the government contribution was relatively small: estimated to UK£3,912,939 as per July 2013 (Gillingham and Buckle, 2014) equivalent to 9.2% of the total budget.	
Complex layered tenure	See "Inclusive" in Table 2.1.	
Delivery of preventive justice	LTR has enabled the provision of evidence of land rights (emphyteutic lease/land certificate, land title).	

(Continued)

TABLE 2.2 (*Continued*)

Assessment of LTR through the Lens of Pro-Poor Land Recordation System Requirements

System Design Elements	Assessment Results: Adherence	Assessment Results: Nonadherence
	Impact on existing land conflicts: results of the study area show an increase of land conflict cases reported in formal courts from 2010 to 2013 with the peak in 2011. Although LTR is not the cause of conflict, its adjudication process has been a trigger for sleeper or given-up cases of land conflict to reemerge.	
	Prevention of land conflict: results from the study area confirm a positive impact of LTR on land conflict. It was found that the number of land conflict cases reported over 5 years after LTR decreased. In addition, it was found that those who have land certificates are likely to report less land conflict.	
Sporadic or systematic implementation	LTR was implemented on a systematic basis and this took relatively few years (5 years).	Up-front investments were little. For instance, regarding the issue of capacity constraints, local land governance institutions running the system had to be addressed up front.
	Awareness raising: LTR used a blend of traditional and innovative communication tools to insure that the public was aware of the program. It was found that the government outreach was highly successful: over 70% attended public meetings (Gillingham and Buckle, 2014). The level of awareness was even higher in places that benefited from special awareness raising programs. For instance, it was found that 79% had attended government meetings on LTR (Santos et al., 2014).	
	Readiness of people for systematic implementation: unfortunately empirical assessment could not be carried out on the readiness of people. Results of the field trial and the baseline study of 2008 indicate that over 80% of the respondents perceive that the impact of LTR would be positive (Sagashya and English, 2010).	

(*Continued*)

TABLE 2.2 (*Continued*)

Assessment of LTR through the Lens of Pro-Poor Land Recordation System Requirements

System Design Elements	Assessment Results: Adherence	Assessment Results: Nonadherence
Flexible spatial index map	Use of high-resolution satellite and aerial imagery was successfully applied by LTR to make the spatial index map.	
Transparency, inclusivity, and equity	Inclusivity and equitable: this suggests right to land recordation without any gender-based discrimination. LTR recorded land with respect to the new land policy and legal framework that provides equal rights to land between men and women: the names of both wife and husband, legally married, were recorded on the land certificate together with the names of the children; female widows' land was recorded as married wife.	Transparency in terms of free and accessible land records: this requirement is not yet met since the newly established land administration structures are still centralized. LTR was also gender biased: illegal wives were not recorded on the land certificate, wives' land in polygamous relationship was recorded but the wife was considered as a friend. Children resulting from illegal relationships, but not recognized by the man, could not be recorded on the land certificate.
Comanagement	Evidence creation and legitimacy of land records: during land demarcation, landowners could identify their boundaries in the presence of their neighbors. Involvement of local community in land recordation: the land committees at the smallest administrative unit (cell) were made using community members. Such committees were in charge of registering and solving any dispute or objection that arose during the adjudication process.	

Source: Data collected by authors 2012–2013.

courts: underlying cases were brought to the surface. Out of 10.3 million parcels that were demarcated and adjudicated, 81% of them were approved for allocation on a lease or title. Some of those that were not approved ended in the courts. Conversely, in areas where LTR was completed, public archives indicate a decrease in land dispute cases reported over 5 years. With regard to equity to access land, LTR enabled equal inheritance right between males and females, co-ownership of land for legally married couples, the requirement to provide consent between legally married men and women in cases of land sales, gifts, or mortgages. Nevertheless, LTR has rendered so-called illegal wives (e.g., those in polygamous relationships) more vulnerable: LTR may have inadvertently weakened the social protections that illegal wives previously enjoyed.

Second, regarding the need for people to legitimize land institutions and vice versa, legitimacy was strengthened by land registration and the issuance of proof of land ownership, also known as legalization. Nevertheless, some categories of people, for example women, were not always considered by social institutions as legitimate inheritors of land. At any rate, newly introduced gender-sensitive land laws have received a high level of perceived legitimacy. It was found that 92% of the respondents believed in equal inheritance right between males and females. However, the effect of this is likely to manifest in the future as younger generation enter the property system. Overall, perceived legitimacy of both social and public institutions is relatively high.

Third, regarding the need for land institutions to be harmonized, a clash between social and public land institutions is apparent: (1) the new public land administration processes introduced through LTR conflict with existing social practices (e.g., land subdivision, land sale/purchase procedures) and (2) gender-sensitive laws previously mentioned, though increasingly gaining people's acceptance, still conflict with existing social practices. Unfortunately, the collected data did not allow for the amount of land conflicts resulting from the new laws to be assessed. This would be worthy follow-up work. Regarding balance and harmony between land policy, land law and other supporting laws, there exist several inconsistencies that are likely to generate land conflicts. For example, Article 10 of the Organic Land law stipulated equal protection of rights over land acquired from both customary and statutory law. In practice, however, this law and other supporting laws (e.g., law regarding Matrimonial Regimes, Family Donations and Succession [Law No 22/99 of 12/11/1999] and the law on Prevention and Punishment of Gender Based Violence [Law 59/2008 of 10/09/2008]) tend to protect the legal wife and children.

Fourth, with respect to people being able to use and enjoy their land rights and restrictions several points can be made. Regarding access to credit, subsequent to LTR a reasonable proportion of the respondents (11%) acquired credit using their land as collateral. There was also evidence of investments into land. The major types of activities included soil conservation measures (anti-erosion structure), reforestation, terracing, and the planting of cash crops. The land market was found to be active, especially the rental market (49% of the respondents were involved in some way), although the private rental markets were not directly managed within the national land administration framework. The study area happened to be in an area reputed to be the most active rural land market nationwide, and results were supportive.

Five years after LTR, the land rental market took over land sales (7% of respondents sold land) and land purchase (18% of respondents purchased land). The latter remains the main mode of land acquisition, after land inheritance, with the mean number of plots acquired through the land purchase being 2.44.

Fifth, regarding the balance between public land restrictions and land rights to be enjoyed, a level of disproportionality was observed. Although some restrictions may be linked to different government policies such as land-use consolidation and land grazing, there exists a bulk of restrictions that originated from LTR: customary individual land ownership was restricted to lease rights. The same restrictions equally apply to the ownership of wetlands, and people were prohibited from subdividing lands below 1 ha. It appears customary land restrictions are not felt, however, it was learnt that family leaders tend to determine and enforce who can buy land: preference tends to be given to members of the family, or any other person known to the family. This is more likely to happen when a female widow needs to sell land. The same occurs when married females have to share land with their brothers at the time of land succession.

Sixth, regarding the need for public institutions to manage and sustain a land information system, LTR led to the creation of a countrywide geographic information system (GIS)-based recordation system. It includes a descriptive database of land claims (or land registry) and spatial details of land parcels. With regard to the proportion of registered land transactions (e.g., land sales, land inheritance, land succession, subdivision or merge of parcels, expropriation), it was found that more effort is needed to ensure rural land transactions are reported to the district land bureau. However, the archives of the office show good evidence of registration of land transactions occurring in urban areas (mainly land sales/purchases and parcel subdivision). This perhaps makes sense as land values are likely to be higher in these areas. The average time to complete a land transaction at district land bureau is set between 15 and 30 business days. In addition, the land transaction process involves other procedures that take place at local level (cell and sector level) and may take an average of 10 business days.

Regarding the alignment of LTR to *fit-for-purpose land administration characteristics* of Enemark et al. (2014), Table 2.1 reveals adherence for many of the elements, except perhaps affordability in the maintenance phase. The government successfully managed to establish the system with help of foreign partners at very low costs to government and citizens. However, it is acknowledged that future financial self-sustainability of the system will be critical. If the poor cannot collect lease documents because of the failure to pay the required fee, it is an indicator that some will likely not be able to afford the land transaction charges required to run the system. The affordability of the new established system is critical and failure to meet this requirement is likely to affect the reliability of the system negatively.

Regarding, the alignment of LTR to *pro-poor land recordation requirements* of Zevenbergen et al. 2013, Table 2.2 reveals the majority were met, although a number were not, and demand specific effort. As per the evaluation against fit-for-purpose characteristics, the affordability of the system for both citizens and state, in the maintenance phase appears to be a potential issue. Meanwhile, with regard to implementation options (sporadic vs. systematic), the readiness of people to accept and use the system appears to be another potential concern. Unfortunately, the data available did not allow a complete analysis in this regard. Finally, recognition of the continuum of land rights was not necessarily adhered to in LTR: a state-prescribed tenure classification was prescribed across the country. Customary tenure arrangements, including those pertaining to illegal wives, tended to be overwritten by the regularized tenure forms.

Discussion

If pro-poor land administration approaches are agreed on as an alternative to providing tenure security for the poor, the evaluation of such approaches should not be limited to their design or establishment phases. In addition, whether the system achieved the original objectives should also be assessed: outputs and outcomes relating to improved land tenure security should be examined. The results presented above enable such an analysis: the following paragraphs consider the inputs, processes, and *also* the outputs of LTR to determine whether land tenure security for the rural poor is provided for.

As discussed in the theoretical background, the frameworks of Enemark et al. (2014) and Zevenbergen et al. (2013) tend to focus on inputs into the land administration system, and to a lesser extent processes. The two frameworks agree on several features: the pro-poor technology options to build the underpinning spatial framework, the need for affordability, and the recognition of a continuum of land rights.

LTR proves that the implementation of pro-poor technology options, and other associated requirements, such as systematic implementation and comanagement is feasible. However, so-called *pro-poor technology solutions* for performing transactions do not appear to be as affordable, if services are being charged at cost price. The case of LTR suggests a significant number of the poor could not afford the initial registration fee: the same group is not likely to be able to afford subsequent land transaction fees. That is, even with pro-poor technology solutions, land administration may remain inaccessible to poorer parts of the population: over time the situation will become unreliable as an evidence base of secure ownership. A potential implication is that following LTR, the poor become more insecure than previously: they do not possess state-backed proof of ownership which was not historically required. Consequently, informal transactions are likely to rise, leading to potential conflicts. Despite such hypotheses, pro-poor technology solutions appear to be able to improve the tenure security of the poor; provided up-front investment by government is undertaken and much emphasis is placed on creating citizen readiness, a requirement of the Zevenbergen et al. (2013) framework.

On recognizing the continuum of land rights, LTR partially accommodates: a continuum of accuracy was observed; however, the continuum of land right options seems to be somewhat passed over. The latter most likely expedited the process, but did result in adverse outcomes for some poorer members of the community (e.g., illegal wives). In the Simbizi et al. (2014) framework, this issue is identified under the relationship dealing with *equity*, however, the Enemark et al. (2014) framework does not deal with the concept explicitly.

The flexible legal framework recommended by Enemark et al. (2014) fit-for-purpose approach is often translated into statutory recognition of customary land rights. In practice, the simple recognition of customary land rights is not enough. The framework of Simbizi et al. (2014) indicates that there should be harmony between social and public land institutions. The case of LTR reveals that the newly introduced legal framework presents areas of inconsistences, both within itself, and via clashes within some existing customary practices. Through initial land recording, both the fit-for-purpose approach of Enemark et al. (2014) and the pro-poor land recordation system of Zevenbergen et al. (2013) give room for a form of land grabbing under the guise of state-based land restrictions. Similarly, a government can use the mechanism of land restrictions and responsibilities to accumulate power over land. The case of LTR arguably illustrates how an imbalance between existing land restrictions and land rights can consequently constitute a source of tenure insecurity.

Overall, the two pro-poor approaches (i.e. Enemark et al. 2014 and Zevenbergen et al. 2013), regardless of how robustly implemented, may not necessarily result in land tenure security for the rural poor. Other key aspects of land tenure security need to be understood and attended to including perceptions of the poor, land institution harmonization, certainty of rights, and balancing land rights with restrictions and responsibilities.

Conclusion

This study commenced by recalling a widely agreed upon argument: conventional land administration systems and associated tools are generally not flexible enough to serve the poor of developing countries. Indeed, this study subscribes to the new mindset: pro-poor land administration approaches and tools are one way of improving delivery land tenure security to the poor. The central inquiry was to examine whether pro-poor approaches to land administration, as prescribed by contemporary literature, provide tenure security to the rural poor. For this purpose, a well-known pro-poor inspired case, the Rwandan LTR, was used to examine how pro-poor land administration works in practice. The program was assessed against three frameworks, all claiming to be pro-poor: fit-for-purpose land administration approach of Enemark et al. (2014), pro-poor land recordation requirements of Zevenbergen et al. (2013), and conceptual model of land tenure security of the rural poor of sub-Saharan African of Simbizi et al. (2014). The three frameworks portray key inputs and outcomes of a pro-poor land administration system, with regard to land tenure security. However, none of them is an end in itself: proposed technical requirements can be implemented but in practice they remain inaccessible to the poor. Furthermore, other suggested requirements are hardly taken into account during project design, and therefore hardly evaluated.

The Rwandan LTR provides a successful implementation story with regard to the fit-for-purpose approach and pro-poor land recordation requirements. Overall, results show that several pro-poor options suggested by the two frameworks can be implemented. Flexibility, in terms of spatial data and land tenure (legal and social) recording, can be achieved. LTR constitutes additional evidence on how a general boundary approach, coupled with the use of affordable technology, such as high-resolution satellite and aerial images, can lead to a flexible land recordation system. The established system is regarded generally reliable, except for a few concerns on the newly introduced wetland boundaries. The LTR process serves as a good lesson when it comes to the level of participation required by the fit-for-purpose approach. However, the replication of the innovative methods used in different context may not lead to the same result in the absence of the strong political will that characterized LTR.

Meanwhile, by excluding existing use rights and the land rights of illegal wives, LTR did not reach the desired level of inclusiveness advocated by the two approaches. Although the implementation process was affordable for government, this was made possible by the support of foreign partners. The so-called pro-poor affordable technology options are still expensive for governments and less affordable for the rural poor, particularly in the maintenance phase: grassroots affordability, a requirement for pro-poor land recordation systems, remains difficult to achieve. To insure the sustainability of the new established system, it is advised to consider the sustainability of the underlying business model of supporting the land administration system: strategies should be developed, preferably during the original project design, that align transactions numbers and income with the costs of maintaining the system (Magis and Zevenbergen 2014). The impact of the affordability issue is that it is likely to undermine tenure security, or may render the poor more vulnerable than prior to LTR. For instance, the acquired perceived tenure security because of land rights recognition is undermined by the land administration services perceived to be physically and financially inaccessible. In addition, the failure to pay the required fee for a land registration certificates may also make the poor more vulnerable. If the poor still cannot afford the so-called pro-poor system, the technology options need to be revisited, or fee subsidies considered again.

At any rate, although it is still too early to convincingly test, the results from the study reveal that LTR is yet to significantly affect several aspects of land tenure security including land market participation, although encouragingly an increase in land investment was observed. At the same time, the limited recognition of the continuum of land rights, the introduction of new land restrictions, and a lack of harmony between new land laws and customary practices will continue to present challenges to land tenure security in the short to medium term.

In conclusion, contemporary pro-poor land administration approaches, as described in the fit-for-purpose and pro-poor land recordation system requirements, can be successfully implemented at national scale. However, it should be recognized that implementation of the elements does not guarantee immediate improved land tenure security for the poor. Other socio-technical elements are crucial to consider including perceptions, equality, and so forth. Pro-poor land administration activities should seek to ensure these aspects are incorporated into project design and assessment. Future work might consider what this entails in terms of approaches to project implementation.

References

Arnot, C. D., Luckert, M. K., and Boxall, P. C. 2011. What is tenure security? Conceptual implications for empirical analysis. *Land Economics*, 87, 297–311.

Benjaminsen, A., Holden, S., Lund, C., and Sjaastad, E. 2009. Formalisation of land rights: Some empirical evidence from Mali, Niger and South Africa. *Land Use Policy*, 26, 28–35.

Bennett, R. M., Van Gils, H. A. M. J., Zevenbergen, J. A., Lemmen, C. H. J., and Wallace, J. 2013. Continuing to bridge the cadastral divide. In: *Proceedings of Annual World Bank Conference on Land and Poverty, April 8–11, 2013*, p. 22. Washington, DC: the World Bank.

Borras, S. M. Jr., and Franco, J. C. 2008. *Land Based Social Relations: Key Features of a Pro-Poor Land Policy*. Oslo Governance Centre Brief 2. Oslo, Norway: United Nations Development Programme, Oslo Governance Centre.

Deininger, K., Ali, A. D., Holden, S., and Zevenbergen, J. 2008. Rural land certification in Ethiopia: Process, initial impact, and implications for other African countries. *World Development*, 36, 1786–1812.

Deininger, K. W. 2003. *Land Policies for Growth and Poverty Reduction*. Oxford, United Kingdom and Washington, DC: Oxford University Press (OUP) and The World Bank.

Enemark, S., Bell, K. C., Lemmen, C. H. J., and Mclaren, R. 2014. *Fit-for-Purpose Land Administration: Open Access e-book*. Copenhagen, Denmark: International Federation of Surveyors (FIG).

FAO. 2002. *Land Tenure and Rural Development*. Rome, Italy: FAO.

Feder, G., and Feeny, D. 1991. Land tenure and property rights: Theory and implication for development policy. *The World Bank Review*, 5, 135–153.

FIG. 1999. Bathurst declaration on land administration for sustainable development: 18th–22th October 1999, Bathurst, Australia. Copenhagen, Denmark: the International Federation of Surveyors (FIG).

Gillingham, P., and Buckle, F. 2014. *Rwanda Land Tenure Regularisation Case Study*. HTSPE Limited for Evidence on Demand, and the UK Department for International Development (DFID).

Government of Rwanda. 2013. *2012 Population and Housing Census. Provisional Results*. Kigali, Rwanda: National Institute of Statistics.

Hagos, H. G. 2012. *Tenure (In) security and Agricultural Investment of Small Holders Farmers in Mozambique*. Maputo, Mozambique: International Food Policy Research Institute.

Independent Evaluation Group (IEG). 2013. *Project Performance Assessment Report: GHANA Land Administration Project*. Washington, DC: the World Bank.

Kumar, R. 2005. *Research Methodology: A Step-by-Step Guide for Beginners*. London, United Kingdom: Sage.

Lavigne-Delville, P. 2006. A conceptual framework for land tenure security, insecurity and securement. *Journal of Land Reform, Land Settlement and Cooperatives*, 2, 24–32.

Lemmen, C. H. J. 2010. *Social Tenure Domain Model: A Pro-Poor Land Tool: e-book*. Copenhagen, Denmark: International Federation of Surveyors (FIG).

Lemmen, C. H. J., Augustinus, C., and van Oosterom, P. J. M. 2009. Social tenure domain model. *Journal GIM International*, 23, 3.

Magis, M., and Zevenbergen, J. A. 2014. Towards sustainable land administration systems: Designing for long-term value creation + powerpoint. In: *Engaging the Challenges, Enhancing the Relevance*, p. 11 and 15 slides. Proceedings of XXV FIG Congress, 16–21 June 2014, Kuala Lumpur, Malaysia.

Payne, G., Durand-Lasserve, A., and Rakodi, C. 2009. *Gender Sensitive and Pro-poor Principles When Regularising Informal Land Settlement in Urban and Peri-urban Area*. Discussion Paper 10. Oslo, Norway: UNDP.

Platteau, J. P. 1996. The evolutionary theory of land rights as applied to sub-Saharan Africa: A critical assessment. *Development and Change*, 27, 29–86.

Sagashya, D., and English, C. 2010. Designing and establishing a land administration in Rwanda: Technical and economical analysis. In: Deininger, K., Augustinus, C., Enemark, S., and Munro-Faure, P. (eds.), *Innovations in Land Rights Recognition, Administration and Governance: Joint Organizational Discussion Paper Issue 2*. Proceedings From the Annual Conference on Land Policy and Administration: Also as e-book. Washington, DC.

Santos, F., Fletschner, D., and Daconto, G. 2014. Enhancing inclusiveness of Rwanda's land tenure regularization program: Insights from early stages of its implementation. *World Development*, 62, 30–41.

Simbizi, M. C. D., Bennett, R. M., and Zevenbergen, J. 2014. Land tenure security: Revisiting and refining the concept for Sub-Saharan Africa's rural poor. *Land Use Policy*, 36, 231–238.

Simbizi, M. C. D., Zevenbergen, J. A., and Bennett, R. M. 2015. Beyond economic outcomes. A pro—poor assessment of land tenure security after regularization in Rwanda. In: Linking land tenure and use for shared prosperity, proceedings of the annual World Bank conference on land and poverty, 23–27 March 2015, Washington DC, United States. 34 p.

Sjaastad, E., and Cousins, B. 2009. Formalisation of land rights in the south. An overview. *Land Use Policy*, 26, 1–9.

Tuladhar, A. M. 2005. Innovative use of remote sensing images for pro-poor land management. *ITC Lustrum Conference: Land Administration: The Path Towards Tenure Security, Poverty Alleviation and Sustainable Development*. Enschede, The Netherlands: ITC.

UN-HABITAT 2007. *How to Develop a Pro-Poor Land Policy. Process, Guide and Lessons*. Nairobi, Kenya: UN-HABITAT.

UN-HABITAT 2008. *Secure Land Rights for All*. Nairobi, Kenya: United Nations Human Settlements Programme (UN-HABITAT).

Van Asperen, P., and Mulolwa, A. 2006. *Improvement of Customary Tenure Security as Pro-poor Tool for Land Development a Zambian Case Study*. 5th FIG Regional Conference: Promoting land Administration and Good Governance, Accra, Ghana.

Van Der Molen, P. 2006. Tenure and tools, two aspects of innovative land administration. In: P. van der Molen and A. Otieno Lamba (eds.), *Decision Makers Meeting: Good Administration of Land: Land Administration for Poverty Reduction and Economic Growth*, Windhoek, Namibia, 7–8 December 2006, p. 23. Windhoek, Enschede, Apeldoorn: School for Land Administration Studies and Polytechnic of Namibia, ITC, Cadastre, Land Registration and Mapping Agency (Kadaster).

Van Der Molen, P., and Lemmen, C. H. J. 2006. Unconventional approaches to land administration: a point of view of land registrars and land surveyors. In: Van Der Molen P. and Lemmen C.H.J. (eds.), *FIG 2005: Secure Land Tenure: New Legal Frameworks and Tools in Asia and the Pacific*, pp. 127–144. Proceedings of an Expert Group Meeting Held by FIG Commission 7, 8–9 December 2005, Bangkok, Thailand. Frederiksberg, Denmark: International Federation of Surveyors (FIG).

Van Der Molen, P., Lemmen, C. H. J., and Meijer, C. 2008. Development and management of services to facilitate pro poor land management and land administration systems. In: *Land Management Information Systems in the Knowledge Economy 2008: Discussion and Guiding Principles for Africa*, pp. 177–196. Addis Ababa, Ethiopia: Economic Commission for Africa (UNECA).

Van Gelder, J. L. 2010. What tenure security? The case for a tripartite view. *Land Use Policy*, 27, 449–456.

Whittal, J. 2014. A new conceptual model for the continuum of land rights. *South African Journal of Geomatics*, 3 (1), 13–32.

Williamson, I., Enemark, S., Wallace, J., and Rajabifard, A. 2010. *Land Administration For Sustainable Development: Also As e-book*. Redlands, CA: ESRI.

Williamson, I., and Ting, L. 2001. Land administration and cadastral trends—A framework for re-engineering. *Computers, Environment and Urban Systems*, 25, 339–366.

Yin, R. K. 2003. *Case Study Research: Design and Methods*. Newbury Park, CA: Sage.

Zevenbergen, J., Augustinus, C., Antonio, D., and Bennett, R. 2013. Pro-poor land administration: Principles for recording the land rights of the underrepresented. *Land Use Policy*, 31, 595–604.

3

Land Administration for Food Security

Rohan M. Bennett, Georgina Rockson, Solomon A. Haile, Joe Nasr, and Liza Groenendijk

CONTENTS

Introduction

Food security is recognized as a key contemporary challenge for the global community (Food and Agriculture Organization [FAO] 2012): projected increases in population suggest smarter food creation and distribution mechanisms are needed at local, national, and global levels (United Nation [UN] 2014; FAO 2014). Assertions in recent years suggest land administration systems support food security: land administration potentially improves access to land, assists in securing land tenures, and subsequently promotes technology investment, which leads to improved agricultural output (Dekker 2001; Deininger 2003; Williamson et al. 2010). Furthermore, land administration systems can support land management activities including land consolidation: in some cases, aggregating fragmented parcels into larger results in more efficient agricultural production (Vitikainen 2004).

Until recently these positive correlations were often assumed by land administration practitioners; however, empirical evidence shows this view might be simplistic (Atwood 1990; Firmin-Sellers and Sellers 1999; Ministry of Foreign Affairs of the Netherlands 2011; Sitko et al. 2014). Empirical proofing is difficult: defining and isolating the appropriate dependent and independent variables is a subjective, debated, and an ongoing task. Despite the debates, new empirical and analytical works emerge. Many are inspired by the proliferation of large-scale agricultural land transactions, sometimes referred to as *land grabbing*, organized by national governments and domestic or foreign investors (Anseeuw et al. 2012). Such deals are argued to threaten the food security of smallholders, but, they also place land tenure and its administration prominently in food security discourse. The volume of new material created, and the disparity of academic domains within which it emerges, makes formation of a holistic and coherent viewpoint difficult. However, this clarity is demanded by policy makers and donor organizations wishing to develop responsible policy interventions (Prakash 2011). Although recent conceptual linkages between land tenure and food security have already been attempted (Ministry of Foreign Affairs of the Netherlands 2011; Bodnar and Hillhorst 2014), the specific role land administration plays, cannot play, and might play requires an updated perspective.

To this end, this chapter seeks to rearticulate the relationship between responsible land administration and food security. Conventional approaches to defining the relationship are generally linear arguments. In response, a systems approach is used here: the network of concepts and relationships existing between land administration and food security will be identified, articulated, and subsequently interwoven. The resultant generalized model is intended to have application for policy makers and land administrators alike: the model acts as a discussion-generating tool to support development of cohesive interventions aimed at supporting the food security endeavors of local communities, non-governmental organizations (NGOs), local and national governments, and international agencies and networks. The remainder of the chapter is structured as follows. First, the theoretical setting is provided: working definitions, key terminology, and existing understandings of interactions are provided. Second, the research methodology consisting of research synthesis, model creation, and evaluation is provided. Third, results from each phase of the research are presented. Fourth, discussion points emerging from the results are provided. The chapter concludes by articulating suggested further work.

Theoretical Perspective

The relationship between food security and land administration is easily discussed in practical terms; however, articulating the relationship in a theoretical setting is more challenging. The lack of consensus on the food security concept lies at the root of this issue. In contemporary popular and political discourses, the term is widely used and its meaning potentially distorted and manipulated, unintentionally or otherwise. However, it was through political channels at the global level, namely the 1974 World Food Conference and the 1996 World Food Summit, that the term finds its most popular definition (UN 1974; FAO 1996). The origins can be further traced to UN's Universal Declaration of Human Rights (UN 1948) and the establishment of the UN FAO in the 1950s (Mechlem 2004). Meanwhile, it finds contemporary relevance as a central concept in the UN's Millennium Development Goals (MDGs), specifically *Goal 1* dealing with hunger and malnutrition reduction. Thus, the definition has evolved over time: issues relating to scale, for example, national versus household level, were dealt with amendments. Attempts have also been made to take into account contextual complexities relating to food availability, access, utilization, and stability (Pinstrup-Andersen 2009). FAO's (1996) current definition for the term states:

> Food security exists when all people, at all times, have physical and economic access to sufficient, safe and nutritious food to meet their dietary needs and food preferences for an active and healthy life.

The four pillars, food availability, food access, food utilization, and food stability, are worth distinguishing. Food availability refers to the supply of food and refers to food creation (or production), net import/export, store amounts, and food aid. It is often considered at the national level. Food access is concerned more with the household and individual levels: adequate national and international food supply does not guarantee it is accessible to all people. Therefore, consideration of incomes, market prices, and expenses are also important. Food utilization is largely about the relationship between the body and the food: ensuring individuals can prepare food, consume food, and improve nutritional status is the focus. Food stability is about the other three dimensions being maintained over time. Food security therefore demands that all four dimensions are simultaneously satisfied (European Commission [EC]-FAO 2008). Moreover, EC-FAO argues it is important to distinguish between two general types of food insecurity: chronic or long term and transitory or short term—each has different causes and remedies.

In the scientific domain, food security is a contemporary crosscutting theme for numerous disciplinary areas: ecology, agricultural science, food science, development studies, social sciences (e.g., economics and sociology), and spatial planning, among others, can all claim legitimate contribution and use. The works of Sulser et al. (2011), Cotula et al. (2009), Allouche (2011), Von Braun (2009), and Zoomers (2010) provide a mere sample—many driven by the increase in large-scale land acquisition activities. Indeed, Hoddinott (1999) suggested, even at that time, there already existed 200 definitions and over 450 indicators of food security. More recently, parts of the disciplinary discourses converged forming the new transdisciplinary study area of *food security* with stand-alone academic journals and fora (Pinstrup-Andersen 2009; Naylor 2011; Magnan et al. 2011). Each discipline adapts and develops the concept to fit with its particular research foci and methodologies. For example, the Centre for Studies in Food Security at Ryerson University, Canada, adds components including *agency*, *adequacy*, and *acceptability* (UN-Habitat/GLTN 2014). Another examples uses: universality, stability, dignity, quantity, and quantity (Mckeown 2006). UN Habitat/GLTN (2014) goes on to provide several other example lenses based on scale, spatial conditions, food-system actors, and vulnerable stakeholders. The lenses of scale and the spatial dimension are also affirmed by Hoddinott (1999). However, a synthesis of all these lenses is not compiled, and at any rate, if one were created it is unclear whether it would support food security research efforts.

In this chapter, the contemporary FAO definition is adopted along with its four pillars: availability, access, utilization, and stability (FAO 2009). Although not a universal definition, it is the most globally recognized, and is therefore an appropriate starting point for defining the contemporary relationship between land administration and food security. A visualized equivalent definition can be found in the works of Riely et al. (1999) and Haeften et al. (2013). These are used in the conceptual modeling work within this chapter.

Meanwhile, the selected definition of land administration finds it origins in UN-ECE (1996):

> The processes of determining, recording and disseminating information about the tenure, value and use of land when implementing land management policies. It is considered to include land registration, cadastral surveying and mapping, fiscal, legal and multi-purpose cadastres and land information systems. In many jurisdictions, land administration is closely related to or facilitates land use planning and valuation/land taxation systems, although it does not include the actual land use planning or land valuation processes.

Despite minor variations in Dale and Mclaughlin (1999), extension by Williamson et al. (2010) to include *land development* as a land administration function (which is also included in this work), and some challenges from Fourie et al. (2002), among others, the definition still carries consensus, if the terms involved are considered in a broad sense: land administration is *generally* considered a subset of land management. At any rate, how best to implement such processes is the subject of far more discussion and debate.

The relationships between land administration and food security definitions are the focus of this work. Rockson et al. (2013) provides a contemporary review in this regard. Building mainly from the work of Dekker (2001), the underlying research synthesis looks at the relationship from two perspectives: (1) how land administration research describes its relationship to food security and (2) how disparate strands of food security research define interactions with land administration. On the second point, the working hypothesis is that land administration was not known to the majority of food security literature: land administration might be implicitly included around discussions on *land tenure*. The review reveals that (1) the amount of literature linking the domains increased during the review period (1995–2012); (2) most studies were observational in nature in that they used comparative methods or research synthesis—design work was less evident; (3) the positive viewpoint, that land administration supports food security, rather than the reverse, dominated—although, negative impacts were certainly demonstrated; and (4) many works, although certainly not all, were not based on strong empirical evidence. The work concludes with a series of generalized statements: a conceptual model is not provided. In this regard, this work seeks to build on the work of Rockson et al. (2013) to formulate such a conceptual understanding.

For such an undertaking, both systems thinking and conceptual modeling are also relevant underpinning theories. As already applied in the land administration domain by Zevenbergen (2002) and Simbizi et al. (2014), systems thinking allows modeling of complex systems through articulation of the dynamic interactions existing between system components (Laszlo and Krippner, 1998). A systematic approach is used to define a system boundary, the relevant entities within, and their interrelationships (Kotiadis and Robinson, 2008). The identification of these boundaries, entities, and relationships is driven by the designer, the investigative process, and the phenomenon of interest itself: the process is necessarily subjective. The result can be visualized using simple modeling tools.

Methodology

To develop the contemporary conceptualization of the links between land administration and food security, the updated *four-pillar* FAO (1996) definition of food security, the related Riely et al. (1999) visualization, and an adapted version of the UN-ECE (1996) land administration, incorporating Williamson et al. (2010) concept, acted as basic foundation frameworks (*c.f.* Rockson, 2012). Then secondary data, originally compiled in the synthesis work of Rockson et al. (2013), the theoretical approach described by Laszlo and Krippner (1998), and the conceptual modeling method of Kotiadis and Robinson (2008), were used to integrate, adapt, and extend the concept by linking the disparate land administration and food security concepts. This involved identification of the system boundary, articulating the existing and new entities, mapping the relationships between those entities, and visualizing the model.

Given that the model is based on data collated by Rockson et al. (2013), it is appropriate to detail the approach used in that study. The synthesis work: (1) considered works published between 1995 and 2012; (2) involved the interrogation of authoritative academic databases including Elsevier Science Direct, Spring Link, Geobase, Sciverse Scopus, and Web of Science; (3) used only keyword, title, and abstract searches; (4) manually recorded in a database qualitative data on a range of food- and land-related criteria, these being developed iteratively throughout the search process; and (5) resulted in 50 papers being selected based on relevance, currency, and degree of relationship to either topics. The process and its limitations should be taken into account when assessing the framework.

To test the validity of the model, the research adopted a quantitative method. The developed conceptual model, together with structured questions, was sent to experts in food or land resource management. The questionnaire sought to validate the conceptual model using experts in the field. The choice of respondents was also critical to the success of the survey. Therefore, the land tenure experts were chosen from the Global Land Tool Network, UN-Habitat, FAO, FIG, Kadaster International (the Netherlands), and the Eastern Africa Land Administration Network (EALAN). Meanwhile, food security experts were identified from FAO, Wageningen University, and the University of Twente (ITC Faculty). The survey was sent to email addresses using Survey Monkey. Aside from the validation process, the survey also sought the opinions of the experts on how the model could be enhanced or improved. On the basis of validation feedback and the recent work of UN-Habitat/GLTN (2014) (completed after the initial study), the model was further refined.

Results

The refined results of the conceptual modeling process are presented in Figure 3.1. The model is not considered context or scale specific: it can apply to community level (including households), or higher levels including state, national, or even the global context. However, given land administration is usually a state or national public-sector activity; it makes most sense considered within these contexts.

The model is best read bottom-up starting from its base; however, a top-down reading is also possible. Four overarching themes provide the basis: resources, land management (including land administration), food governance, and food security. Each theme includes numerous entities or concepts (rounded boxes): the relationships between these concepts illustrate the hypothesized relationship between the themes (the arrows)—and ultimately between land administration and food security. In any context, these might be reinforcing or undermining relationships. The shading of themes conveys no meaning, although, it could be considered that lighter shading indicates a less tangible theme and related entities: ones that are more difficult to articulate and measure.

At the base of the diagram sits the *resources* theme; consisting of built resources, natural resources, capital resources, and social resources. The resources existing within have a strong determining impact on food security. Generally speaking, contexts equipped with high resource levels are more likely to achieve food security than those areas where resources are unavailable or depleted. As a reminder of this, "un" is added to the theme description. These four resource categories differ slightly from the Riely et al. (1999)

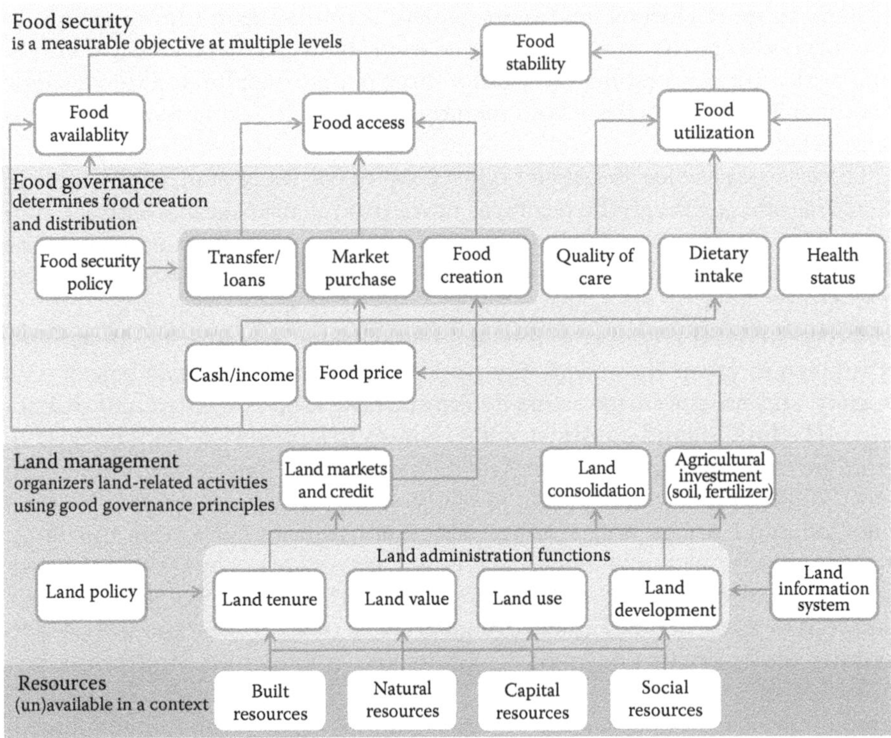

FIGURE 3.1
Land administration for food security. (Adapted and extended from Rockson, G., *Land Administration for Food Security: A Research Synthesis*, ITC, University of Twente, Enschede, The Netherlands, 2012.)

model: (1) Riely's *community resources* and *human resources* are combined under *social resources*—both deal with human capacity, or social capital, and there was no need to distinguish in the model and (2) *built resources* is added to reflect the importance of the context's transport and service infrastructures (e.g., internet, electricity) in determining food security outcomes.

The *land management* theme sits atop resources, and is largely a replication of the Williamsonet al.'s (2010) land management paradigm. More recently aligned with the principles of good governance, the paradigm argues that a country's (or state or city) context (i.e., the resources) are managed (or mismanaged) by land administration functions (i.e., those concepts in the more lightly shaded box) relating to land tenure, land value, land-use planning, and land development—although, obviously these functions focus mainly on the management of built and natural resources. Further, the sound design and implementation of the functions is argued to be dependent on an agreed land policy and accompanying land information systems (digital, or not). The functioning land policy, land administration system, and land information system, support various land management activities. Important in the context of food security are land consolidation, land markets, credit markets, and agricultural investment (the latter three being consolidated into a single concept): it is through these land management activities that food security is supported (or not).

The *food governance* theme embodies most of the Riely et al. (1999) conceptual diagram, although the term was never used in association with the conceptual model. No standardized definition exists—although its use can be found in sociology and development studies (Lang 2003; Oosterveer 2005). The term "global governance of food security" is more commonly used in UN discourse. At any rate, in this diagram, similar to *governance* in general, it is taken to mean those activities undertaken by various state-based, civil society, and private sector actors determine how food is created and distributed. The link to land administration comes via land consolidation, agricultural investments, and land markets (Grossman and Brussaard 1989; Jansen et al. 2010). How these activities are performed has a consequence for food creation, food prices, and also arguably cash/income (although, this latter relationship is not shown here). That is, empirical works show that land consolidation, investment in soil and fertilizer, and formation of land markets can lead to lower food prices and higher levels of food creation: the need to acquire food transfers or loans is diminished. Meanwhile, food prices and available income also determine whether local, national, or international markets are used. Another link to land administration, although not always implemented, is via policy: land policies are implicitly, and should be explicitly, linked to food security policy. Food security policy will determine the local/city, state, or national stance with regard to *food loans, markets purchase,* or self *food creation*. These appear more relevant to the state or national level, although the role of the local or city level should not be underestimated. The final three concepts are more disconnected from the others: *quality of care, dietary intake,* and *health status*. Although relationships to land administration can be drawn, these concepts are more related to household health levels and capacity—and are considered more directly related to cash/income level (although, a link to social resources should also be made).

The top theme, food security is considered the overarching objective, more than anything else—and consists of the four UN (1996) pillars: *food availability, food access, food utilization,* and *food stability*. It links to the other themes via food governance. Specifically, food availability is determined by food creation, *cash income,* and *food prices*. Meanwhile, food access is determined by *transfers/loans, market purchase,* and/or food creation. Finally, food utilization is usually considered at the household level and therefore relates to household indicators including the quality of care regarding food resources, dietary intake, and health status. As already mentioned, food stability is a product of the other three pillars occurring over a long period.

The conceptual model (Figure 3.1) is the result of refinements to an earlier version, against which exploratory evaluations were made during the preliminary testing and validation process. The online questionnaire provided participants with a simplified model (only including the four themes) and a more complex version (similar to Figure 3.1). To avoid confusion, only the refined model is presented in this chapter; however, the synthesized results from the validation of the simplified and complex model are presented to highlight key concerns and issues as identified in the questionnaire (Table 3.1). In total, 25 participants responded to the survey. Regarding disciplinary backgrounds, 21 of the participants identified as land experts, and only 4 identified themselves as food experts. Some participants also described themselves as experts in either remote sensing or natural resource management. The limitations and bias of the sample group should be kept in mind when interpreting the results.

The questionnaire revealed a range of issues with the original model. First, regarding understanding the simplified thematic model, the majority of respondents sensed they fully understood the model, or at least partially understood. To facilitate answers, a range of generalized responses were made available to participants. The most specific issues raised by participants were that (1) the model was too simple and general to be useful, (2) other factors determine food governance, (3) food security was only managed at governmental level, and (4) some participants wanted more explanation of food governance and resources.

Second, regarding agreement with the simplified thematic model, the majority of respondents partially agreed. The most specific issues raised in support of answers were that (1) external factors such as harsh climatic conditions affect food security negatively and must be considered; (2) food governance should be changed to farm governance whereas land management

TABLE 3.1

Results from Exploratory Validation

	Yes	No	Partial	Others
1. Do you understand the model presented above? Referred to simplified model	16	3	6	
2. Do you agree with the model presented above? Referred to simplified model	9	2	14	
3. Do you understand the model presented? Referred to detailed model	13	3	9	
4. Do you agree with the model presented? Referred to detailed model	15	10		
5. How might the model presented be enhanced or improved?				14 answered, 11 skipped
6. Please identify the field to which you most strongly identify?				21 land expert, 4 food expert

should also be changed to resource management with more explanation; (3) food security depends largely on technology and processing of perishables for future use, and this was not evident in the model; and (4) there was confusion on flow of arrows on the resources boxes. On the final point, arguments could be made for directionality to flow both ways; alternatively, arrowheads could be removed altogether.

Third, regarding understanding of the detailed model, roughly half of the participants suggested understanding. The most specific issues raised were that (1) the arrow linking land management to food governance should end on the outer side of the food governance system, (2) cash/income directly affects dietary intake, (3) food utilization also affects health status hence there should be two sided arrows, (4) renaming the model *food land management* should be considered, (5) a better explanation on the difference between community resources and *natural resources* should be given, (6) food utilization should be better explained as part of food governance and a definition of loan or transfer should be provided, and (7) explanations for the reasons for highlighting certain areas as food security impediments should be provided.

Fourth, regarding agreement with the detailed model, the majority of the participants agreed with the model. The most specific issues raised were that (1) there should be policy interventions as regards food governance as it is for land governance; (2) *land policy* should be made explicit and broader than the figure; (3) resources could be a major impediment as well, therefore an arrow should link directly from resources to food security; and (4) system boundaries are basically missing and subsystems are not defined.

Fifth, regarding how the model might be improved, only 14 responses were provided to this question. The specific requests were (1) to make clear whether land development (presumably including land consolidation or readjustment) is considered in the food creation processes; (2) to start with *good governance* first followed by the land administration paradigm; (3) to include economic, social, and environmental developments; and (4) to consider severity of food insecurity classified by Integrated Food Security Phase Classification (IPC) as a tool to measure food security severity to aid in emergency response and humanitarian efforts, as used in Somalia and Kenya.

Discussion

Discussion is undertaken using the following thematic areas: model acceptance, system boundary, entities, and relationships. Crosscutting issues are discussed in each theme. These include the modeling of space and time dimensions and the strengths and weaknesses of the approach and modeling decisions made.

Regarding acceptance of the model, as articulated in UN-Habitat/GLTN (2014), food security is broad and multifaceted concept. It can be considered a process, or an outcome. It resonates at multiple levels of governance, geographic scale, and unit of analysis. It is also an area of inquiry, with multiple and disparate lenses being applied to the concept. Similarly, *land administration* exhibits similar characteristics, although, perhaps to a lesser extent. Therefore, using a limited literature synthesis, in an effort to capture all the involved concepts and interrelationships, will necessarily result in an incomplete model—at least from the perspective of some. As mentioned in the introduction, the model developed here was aimed primarily at high-level decision and policy makers, and to a lesser extent at those practitioners working in the land administration and food security fields. As such, these stakeholders were targeted in the validation process—and pleasingly, among this group, the model was understood and acceptable to the majority.

At any rate, the assignment of the systems boundary remains challenging. Via the Williamson's et al. (2010) land management paradigm, good governance concepts frame the model: inputs, processes, and outputs relating to land and food were major starting points. In addition, the four thematic areas act as subsystems having their own boundaries. The approach has implications for scale of applicability and also for the very nature of the model itself. For example, regarding scale, the original Riely et al. (1999) food security model was nested in social, policy, and natural *environments,* and appears oriented to the national level. Similarly, Williamson's et al. (2010) model explicitly works from the *country context* into more specific concepts. However, in the model presented here, an attempt was made to neutralize the level of application by using the term "context" instead: it is suggested the model and concepts included are potentially applicable at the local, state, national, or partially even international level. However, this sentiment should be treated carefully: the model only acts as a starting point in this regard. Modifications and adaptations would be needed for specific scales and contexts. Meanwhile, regarding the nature of the system itself, it represents not only the relationships between concepts, but also the dynamic process of nested food and land-related inputs and outputs. The nature of the inputs and outputs varies: there are more tangible *things* (e.g., natural resources), but also subprocesses (e.g., land consolidation), and also what appear to be indicators (e.g., food access) that may or may not be easily measured. This flexibility is considered a strong point, but, potentially makes model interpretation difficult. At any rate, the flexibility still results in several of the lenses mentioned in UN-Habitat/GLTN (2014) being potentially missed or downplayed in the model.

Regarding the selected and articulated entities, as already mentioned, land-related and food security literatures with an emphasis on land, informed the selection of entities. The validation revealed that although the broad definitions of land administration and food security carried consensus among the group, sub-concept or entity definitions (e.g., community vs. natural resources or natural resource management vs. land management) were not always known or shared. In addition, several concepts were suggested to be overlapping (e.g., land management and food creation). For this reason, any model attempting to link food security with land administration requires textual accompaniment (as provided in Results): the shared language of the collected stakeholder group is not standardized and explanations are needed. The Results section in this chapter attempts to provide this needed clarity. Meanwhile, some suggested that other important factors were not represented (e.g., dynamic climatic factors). With these limitations in mind, it can be said that the model is most likely not adequate for specific lines of land/food research. For example, it is not fine grained enough to explicitly reveal relationships between land administration and food, relating to gender, customary groups, informal rental markets, and potentially even large-scale land acquisitions. These issues are hidden, for example under the broad land tenure land administration function. Even the current and potential role of technology is hidden for example under the concepts of food creation and land development—although *land information system* is explicitly mentioned. Significantly, for the intended audience, land and food security are explicit concepts—the latter being a refinement on the original model presented to the validation group. One entity that appears to be missing in the model is *population*. Population change is an important factor as it determines resource scarcity, and therefore food scarcity; however, the models and works studied do not seem to deal with this issue—and its impact on the food security situation.

With respect to the relationships between the entities, these represent the most important part of the model: through the entities, they conceptually tie the whole model together. Some examples include: *land policy* links to *food security policy*; *land consolidation* links to *food creation* and also links to *food prices*; and *land markets/credit/agricultural investment* link to *food prices* and also link to *food creation*. In the desired scenario, these links are positive: the land-related entities support food security. In an undesirable scenario, no land policies exist, or are not aligned with food security policy; land is highly fragmented; land tenures are not secured; land and credit markets do not exist, or fail; and therefore agricultural investment is undermined. Further, on the model's relationships, the validation revealed varied perspectives on several of them. The broad nature of several of the entities means it would be quite possible to draw direct relationships between most entities. However, this would undermine the utility of the model, as such only those relationships considered of highest importance were depicted in the model. At any rate, in some cases, the advice of the validation responses was acted upon. For example, *cash income* was not originally linked to *dietary impact* but now is. A limitation of the represented relationships is that they are not labeled in the diagram (although, they are described in the textual accompaniment). The aesthetic of the diagram is the main reason for the noninclusion: the depiction would become cluttered and unreadable otherwise. In addition, all arrows only flow in one direction. Again, this was primarily to enhance readability of the diagram—and promote the idea that it is the four pillars of food security that are ultimately aimed at—as per the original Riely et al. (1999) and the Williamson et al. (2010) models. However, it is conceded that several arrows could flow in both directions.

Conclusions

This chapter recognized that food security is a key contemporary challenge for the global community and that land administration systems are often considered among the support systems that lead to positive food security status. Empirical proofing of the relationships has been difficult: defining and isolating the appropriate dependent and independent variables remains an ongoing task. Therefore, the aim of the chapter was to conceptually model the specific role land administration plays, cannot play, and might play in supporting food security.

To develop the contemporary conceptualization, the updated FAO (1996) definition of food security and an adapted version of the UN-ECE (1996) land administration, incorporating Williamson et al. (2010) and good governance concepts, acted as basic foundation frameworks. Then, secondary data—originally compiled in the synthesis work of Rockson et al. (2013), the theoretical approach described by Laszlo and Krippner (1998), and the conceptual modeling method of Kotiadis and Robinson (2008)—were used to integrate, adapt, and extend the conceptual links between the disparate land administration and food security concepts. To test the validity of the model, a structured questionnaire was sent to knowledgeable professionals who have expertise in food or land resource management.

The model consists of four overarching themes: resources, land management (including land administration), food governance, and food security. Each theme includes numerous entities or concepts (rounded boxes): the relationships between these concepts illustrate the hypothesized relationship between the themes (the arrows)—and ultimately between land administration and food security. The key entities were found to be food security policy, land policy, land consolidation activities (via land tenure), land markets and credit, agricultural investment, food creation, and food price. These interacted through multiple relationships. The model attempts to be neutral in terms of geographic scale, although in this regard application in specific scales would require adaptation. Finally, the model possesses a dynamic quality in that it represents processes.

The majority of respondents sensed they understood and agreed with both the themes and the detailed aspects of the model. However, specific concerns were raised with the definition of the system boundary, selection of entities, and the relationships. These suggestions were in some cases incorporated into the revised design.

The model developed was aimed primarily at policy makers and practitioners in both land administration and food sectors. It sought to provide a simple mapping of concepts that reveals the significance of land administration activities in food security interventions and vice versa. Consequently, it reveals the potential for land administration activities to undermine or support food security and insecurity. In this way the model can be used to provide options to support cohesive development of an intervention strategy consisting of multiple activities. Beyond this, the model can be used as basis for training programs and further research on specific relationships. However, the model has several caveats and limitations: it is based on a limited research that purposely focused on specific terminology from the land administration domain—related concepts from other domains were not intensively studied (e.g., land tenure in agricultural studies); it only focuses on relationships between core land administration activities and central food security pillars; it is simplistic in that it is not fine grained enough to explicitly reveal relationships between land administration and food; it requires textual accompaniment; and it lacks a strong spatial component. In this regard, further work could also consider the *right to food*, and whether it possesses a spatial quality, and further validate the model via a broader expert survey.

References

Allouche, J. 2011. The sustainability and resilience of global water and food systems: Political analysis of the interplay between security, resource scarcity, political systems and global trade. *Food Policy* 36: S3–S8.

Anseeuw, W., M. Boche, T. Breu, M. Giger, J. Lay, P. Messerli, and K. Nolte. 2012. *Transnational Land Deals for Agriculture in the Global South: Analytical Report Based on the Land Matrix Database*. The Land Matrix Partnership. Available at http://www.landcoalition.org/en/publications/transnational-land-deals-agriculture-global-south.

Atwood, D.A. 1990. Land registration in Africa: The impact on agricultural production. *World Development* 18(5): 659–671.

Bodnar, F., and T. Hilhorst. 2014. Impact pathways land tenure food security. In: *Draft Report on the Interim Workshop for Land Tools for Food Security Project*. The Hague, Netherlands.

Cotula, L., Vermeulen, S., Leonard, R. and Keeley, J., 2009. *Land Grab or Development Opportunity? Agricultural Investment and International Land Deals in Africa*. London, United Kingdom: International Institute for Environment and Development.

Dale, P., and J. McLaughlin. 1999. *Land Administration*. Oxford, United Kingdom: Oxford University Press.

Deininger, K.W. 2003. *Land Policies for Growth and Poverty Reduction*. Oxford, United Kingdom and Washington, DC: Oxford University Press (OUP) and World Bank Publications.

Dekker, H.A.L. 2001. A New Property Regime in Kyrgyzstan: An Investigation into the Links between Land Reform, Food Security, and Economic Development. PhD Thesis. Amsterdam, The Netherlands: University of Amsterdam.

EC-FAO. 2008. *An Introduction to the Basic Concepts of Food Security. Food Security Information for Action: Practical Guidelines*. Available at http://www.fao.org/docrep/013/al936e/al936e00.pdf.

FAO. 1996. *Rome Declaration on World Food Security*. Rome, Italy: FAO.

FAO. 2009. *Declaration of the World Food Summit on Food Security*. Rome, Italy: United Nations.

FAO. 2012. *Voluntary Guidelines on the Responsible Governance of Tenure of Land, Fisheries and Forests in the Context of National Food Security*. Rome, Italy: CFS and FAO.

FAO. 2014. *The State of Food Insecurity in the World 2014. Strengthening the Enabling Environment for Food Security and Nutrition*. Rome, Italy: FAO.

Firmin-Sellers, K., and P. Sellers. 1999. Expected failures and unexpected successes of land titling in Africa. *World Development* 2(7): 1115–1128.

Fourie, C., P. Van der Molen, and R. Groot. 2002. Land management, land administration and geospatial data: exploring the conceptual linkages in the developing world. *Geomatica* 56(4): 351–361.

Grossman, M., and W. Brussaard. 1989. *Reallocation of Agricultural Land under the Land Development Law in the Netherlands*. Wageningen, The Netherlands: Wageningen Universiteit.

Haeften (van), R., M.A. Anderson, H. Caudill, and E. Kilmartin. 2013. *Second Food Aid and Food Security Assessment (FAFSA-2) Summary*. Washington, DC: FHI 360/FANTA.

Hoddinott, J. 1999. *Operationalizing Household Food Security in Development Projects: An Introduction*. Washington, DC: International Food Policy Research Institute.

Jansen, L.J.M., M. Karatas, G. Kusek, C. Lemmen, and R. Wouters, 2010. The Computerised Land Re-Allotment Process in Turkey and The Netherlands in Multi-Purpose Land Consolidation Projects, 2010 FIG Congress, Sydney, Australia.

Kotiadis, K., and S. Robinson. 2008. Conceptual modeling: Knowledge acquisition and model abstraction. In *Proceedings of the 40th Conference on Winter Simulation*, pp. 951–958. Winter Simulation Conference. Austin, TX: IEEE.

Lang, T. 2003. Food industrialisation and food power: Implications for food governance. *Development Policy Review* 21(5–6): 555–568.

Laszlo, A., and S. Krippner. 1998. Systems theories: Their origins, foundations, and development. In: *Systems Theories and A Priori Aspects of Perception, Advances in Psychology*, edited by J. C. Jordan, pp. 47–74. Amsterdam, North-Holland: Elsevier.

Magnan, N., T.J. Lybbert, A.F. McCalla, and J.A. Lampietti. 2011. Modeling the limitations and implicit costs of cereal self-sufficiency: The case of Morocco. *Food Security* 3(1): 49–60.

McKeown, D. 2006. *Food Security: Implications for the Early Years*. Background paper. Ontario, Canada: City of Toronto Public Health. Available at https://www1.toronto.ca/city_of_toronto/toronto_public_health/healthy_public_policy/children/files/pdf/fsbp_exec_summary.pdf.

Mechlem, K. 2004. Food security and the right to food in the discourse of the United Nations. *European Law Journal* 10(5): 631–648.

Ministry of Foreign Affairs of the Netherlands (MFAN). 2011. Improving food security: A systematic review of the impact of interventions in agricultural production, value chains, market regulation, and land security, IOB Study, no. 363, December.

Naylor, R. 2011. Expanding the boundaries of agricultural development. *Food Security* 3(2): 233–251.

Oosterveer, P. 2005. *Global Food Governance*. Wageningen, The Netherlands: Wageningen University.

Pinstrup-Andersen, P. 2009. Food security: Definition and measurement. *Food Security* 1(1): 5–7.

Prakash, A., ed. 2011. *Safeguarding Food Security in Volatile Global Markets*. Rome, Italy: Food and Agriculture Organization of the United Nations.

Riely F., N. Mock., B. Cogill, L. Bailey, and E. Kenefick. 1999. *Food Security Indicators and Framework for Use in the Monitoring and Evaluation of Food Aid Programs*. Arlington, VA: Food Security and Nutrition Monitoring Project (IMPACT), ISTI.

Rockson, G. 2012. *Land Administration for Food Security: A Research Synthesis*. MSc Thesis. Enschede, The Netherlands: ITC, University of Twente.

Rockson, G., R.M. Bennett, and L. Groenendijk. 2013. Land administration for food security: A research synthesis. *Land Use Policy* 32: 337–342.

Simbizi, M.C.D., R.M. Bennett, and J. Zevenbergen. 2014. Land tenure security: Revisiting and refining the concept for Sub-Saharan Africa's rural poor. *Land Use Policy* 36: 231–238.

Sitko, N.J., J. Chamberlin, and M. Hichaambwa. 2014. Does smallholder land titling facilitate agricultural growth? An analysis of the determinants and effects of smallholder land titling in Zambia. *World Development* 64: 791–802.

Sulser, T.B., B. Nestorova, M.W. Rosegrant, and T. van Rheenen. 2011. The future role of agriculture in the Arab region's food security. *Food Security* 3(1): 23–48.

UN. 1948. *Universal Declaration of Human Rights*. New York: United Nations. Available at http://www.un.org/en/documents/udhr/index.shtml.

UN. 1974. *Universal Declaration on the Eradication of Hunger and Malnutrition*. New York: Office of the High Commissioner for Human Rights. Available at http://www.ohchr.org/EN/ProfessionalInterest/Pages/EradicationOfHungerAndMalnutrition.aspx.

UN. 2014. *The World Population Situation in 2014: A Concise Report*. ST/ESA/SER.A/354. New York: Department of Economic and Social Affairs Population Division.

UN-ECE. 1996. *Land Administration Guidelines*, p. 111. ECE/HBP/96. Geneva, Switzerland: UN Economic Commission for Europe.

UN-Habitat/GLTN. 2014. *Discussion Paper for Research Project on Land Tools for Food Security*. Nairobi, Kenya: Global Land Tools Network, UN-Habitat.

Vitikainen, A. 2004. An overview of land consolidation in Europe. Nordic Journal of Surveying and Real Estate Research 1(1): 25–44.

Von Braun, J., and R. Meinzen-Dick. 2009. *Land Grabbing by Foreign Investors in Developing Countries: Risks and Opportunities*. Washington, DC: International Food Policy Research Institute.

Williamson, I., S. Enemark, J. Wallace, and A. Rajabifard. 2010. *Land Administration for Sustainable Development*. Redlands, CA: ESRI Press Academic.

Zevenbergen, J. 2002. *Systems of Land Registration Aspects and Effects*. PhD Thesis. Delft, The Netherlands: TU Delft.

Zoomers, A. 2010. Globalisation and the foreignisation of space: Seven processes driving the current global land grab. *Journal of Peasant Studies* 37(2): 429–447.

4

Urbanization, Land Administration, and Good-Enough Governance

Berhanu K. Alemie, Rohan M. Bennett, and Jaap Zevenbergen

CONTENTS

Introduction

The growth of cities, through the process of urbanization, delivers both positive and negative outcomes. Cities are engines for economic and technological transformation; however, they are also the places where slums, informal sectors, inefficient urban services, and conflict over land resources manifest (UN-Habitat 2012). The duality of outcomes draws attention to improving urban land governance (Dobson et al. 2014).

Governance is argued as both a cause of and a solution for the challenges inherent to contemporary urban environments, the former because many urban challenges are the result of weak governance (Szeftel 2000) and the latter because the challenges can be addressed through the so-called good governance, involving informed and transparent decision making, that results in prosperous and equitable cities (UN-Habitat 2012). In this chapter, land governance refers to the following:

> The rules, processes and structures through which decisions are made about access to land and its use, the manner in which the decisions are implemented and enforced, [and] the way that competing interests in land are managed (Palmer et al. 2009).

Although the concepts of "governance" and even "land governance" are adequately covered in the literature, little attention is given to the specific case of urban land governance. There are various reasons that support the separate study of urban land governance from land governance more generally. In some countries, Ethiopia, for example, there are separate land policies and laws for urban and rural lands: urban and rural lands are administered by distinct institutions and organizations. In addition, urban and rural settings have different physical and social dynamicity. Urban areas are usually more dynamic both physically and socially compared with rural areas. The magnitude and consequences of poor land management is ultimately different in urban and rural areas, and the spatial accuracies of information required for decision making in rural and urban areas are also different. Furthermore, urban environments, in their current new sizes, for example, megacities of over 10 million inhabitants, are quite unprecedented in human history, particularly with respect to the number that will exist globally. The process of urbanization is creating new environments with new challenges, challenges relating to land provision, housing, food and water security, infrastructure provision, waste management, and transport/movement. The vertical growth of cities along with their temporal dimension also adds further need for enquiry into information on three-dimensional/four-dimensional properties. The need to independently study the urbanization of land and urban lands more generally is clear.

Governing urban land, and the urbanization process, is about dealing with urban people to urban land relationships. Information regarding the people to land relationships such as descriptions of ownership, types of land rights, values, and uses is pertinent in this respect. In other words, the support of cadastres, land registers, and administration systems is crucial. In this chapter, the definition by Williamson et al. (2010) of cadastre is used:

> Parcel based and up-to-date land information system containing a record of interests in land (i.e., rights, restrictions, and responsibilities). It usually includes a geometric description of land parcels linked to other records describing the nature of the interests, the ownership or control of those interests, and often the value of the parcel and its improvements.

That is, the cadastre is considered as a system comprising both the cadastral map and the land register. The same is applied by Bogaerts and Zevenbergen (2001) and Silva (2005). Land registration here is considered the part of the cadastral system that incorporates

> The process of recording land ownership, rights to land, and obligations of land owners and users (van der Molen 2011).

Previous works affirm the central support cadastral information can play in sustainable urban development (Williamson et al. 2010; Bennett et al. 2012; Zevenbergen et al. 2013), economic development (de Soto 2000), environmental protection (Guo et al. 2013), and land governance (UN-FIG 1999).

However, Barry and Fourie (2002) argue that the cadastre can also impede development. Undesired outcomes of cadastral implementation are common, especially in developing countries (Obeng-Odoom 2012), including Ethiopia (Alemie et al. 2015a).

Meanwhile, most representations or conceptual models that attempt to link cadastres with urban land governance appear to have at least one of three limitations. First, they tend to focus on describing either a positive or a negative viewpoint: the range of potential outcomes is not displayed. For example, the "Land Management Paradigm" (Enemark 2005) only depicts the positive relationship: scenarios of unsuccessful and undesired outcomes of a cadastral implementation on land management are not considered. However, as discussed earlier, cadastres may be detrimental to good land governance in some cases. Thus, conceptual models should also portray the pitfalls that lead to the possibility of unsuccessful outcomes. A more balanced, if not holistic, representation would provide decision makers and practitioners with a multidimensional understanding. Second, many models do not represent the importance of spatial dimensions in terms of land governance inputs, processes, and outcomes. The natural and built environments have a significant influence on all these elements. Third, they tend to be linear in nature: they do not provide for understandings of land rights and tenure security as a continuum (GLTN 2008), constellation (Benda-Beckmann et al. 2006), or web of interests (Arnold 2002). Arguably, land governance actors and processes, where the issues of land rights and tenure security are embedded, can also be viewed as a continuum of different types of land governance. However, existing models lack a comprehensive characterization of the types of land governance across the land governance continuum. Therefore, a conceptual model that fills these gaps is required to improve understandings of urban land governance.

In response to the aforementioned issues, the aim of this chapter is to develop a conceptual model that (1) is more neutral on positive and negative linkages between cadastres and urban land governance; (2) is more inclusive of the spatial component; and (3) demonstrates the types of urban land governance across a continuum by integrating the inputs, processes, cadastral influences, and spatial outputs of land governance. First, the underlying theoretical perspective is further described. A description of the research methodology follows. The results, along with the developed conceptual model, are then presented. Discussion follows, and major conclusions regarding the conceptual model are forwarded.

Theoretical Perspective

This section discusses theoretical concepts that underlie urban land governance and debates relating to the role of urban cadastres.

Urban land governance, as a concept, grows in significance due to the rapidity of contemporary urbanization processes. Urbanization increases the demand for land: more land users and land interests are involved than in rural areas. The new interests can put enormous stress on land (Thuo 2013). Well-organized decision-making processes regarding urban land are vital. However, this is often a challenge in developing countries: the new actors and interests are diverse and not easily harmonized (Ligtenberg et al. 2009); weak institutions and high land values mean land is the focus of corrupt actions (Burns and Dalrymple 2008); the rapid and often unplanned illegal conversion of rural lands to urban lands leaves many actors out of the decision-making process, for example, in Ethiopia (Melesse 2005); urban land laws are often subjected to constant change, meaning actors are misinformed, confused, or untrusting (e.g., in Ethiopia the urban land leasehold proclamation was modified three times since being incepted in 1993, and in China land policies have changed dramatically since 1949, Gao et al. 2014); the growth of cities and associated land demands are supported by obsolete spatial plans (Dawson et al. 2014; Fekade 2000); the laws are usually formulated and implemented without an underlying policy (Africa Union Commission 2010); the institutional and organizational functions that are responsible for dealing with issues of people to land relationships are weak and fragmented (Williamson et al. 2010); and cadastres are not pro-poor and may serve only more elite social groups (Zevenbergen et al. 2013). These issues combined deter policy implementation and decision making and consequently lead to undesirable outcomes.

McNeill et al. (2014) suggest that incorporating governance concepts into policy making and implementation is central for tackling the aforementioned challenges: applying a governance concept creates a platform that encourages different actors to participate, where various interests are discussed and argued, and collaboration during policy making is generally strengthened. In the context of urban land, for example, incorporating governance concepts has at least three advantages. First, it assists in identifying the root causes of urban people to urban land problems such as urban land access and land use at the local level, where the epicenter of urban development is located (Rakodi 2003). Second, a governance approach provokes discussions among the diverse actors, including urban people, that enable scrutiny of alternative solutions to the problems identified. Third, it forms a shared platform that allows for follow-up implementation of identified solutions. These factors combined can lead to the achievement of the desired policy outcomes, and thereby the goals of sustainable urban development.

Land governance benefits broader public governance (Burns and Dalrymple 2008; FAO 2007) especially in urban areas. This is because contemporary urbanization and associated public governance problems such as provision of housing, utilities, infrastructures, and waste management have urban land dimensions and obviously can be dealt with via the notion of urban land governance. Solving these problems can improve the lives of the urban poor and consequently supports the realization of sustainable development in a country or nation more broadly (Bennett and Alemie 2015; Williamson et al. 2010).

Urban land governance as a continuum represents one conceptualization of land governance. The idea is inspired by the continuum of land rights: the range of possible forms of land tenure is increasingly considered a continuum (GLTN 2008). In the context of land governance, it refers to the forms or types of land governance that can exist during urban people to urban land relationships. Across this continuum, three broad types of urban land governance are considered in this chapter. These include good urban land governance, good-enough urban land governance, and bad urban land governance.

In the view of Grindle (2011), good governance is the type of governance that could tackle all the problems that emanate from institutional, actor, and political constraints at once. However, this is perhaps more a vision than a reality: achieving good governance in "one hit" is difficult, if not impossible. This is well described by Foster (2000), who suggests that "good governance is easy to talk about but hard to do": it is difficult to achieve good governance at the operational level. According to Grindle (2011), attempting to resolve all governance issues and create good governance is a waste of time and resources.

The second type of urban land governance in the continuum is good-enough urban land governance. It considers an intermediate set of options, based on the basic needs of the society, at a given time and socioeconomic condition. The good-enough ideology is flexible to future upgrading when and if the appropriate resources and capacities prevail. In this sense, good-enough land governance is similar to the description of "intermediate tenure options" (Payne 2005) or "fit-for-purpose" land administration (FIG 2014). For example, prioritizing the establishment of a legal land right, and guaranteeing land tenure security, can be an option to support emerging market economies that lack resources and skills to establish a "good" system, in all its aspects. The third type of governance in the continuum is bad governance. This refers to a situation where neither the good nor the good-enough governance objectives are attempted or achieved.

The three types of urban land governance are described by inputs and processes and constitute policies; laws and diverse actors; cadastres of different quality; and resultant outputs indicators, such as informal settlements. The inputs and processes determine the nature of urban land governance outputs that can be easily understood through spatial analysis (Alemie et al. 2015b). For example, in a good governance situation policies and laws exist and are appropriately designed and implemented by the participation and collaboration of diverse actors, and cadastres are efficient and support the achievement of good quality output indicators. In good-enough urban land governance, the formulation and implementation of policies and laws, actor participation and collaboration, and influence of the cadastre combine to support the achievement of an intermediate quality of output indicators. In bad governance, policies and laws may or may not exist; actors are very few and do not collaborate; and cadastres are poor, perhaps even playing an undermining role in the community. The situation results in bad quality of output indicators. Across this land governance continuum, spatial analysis can enhance the understanding of the spatial manifestations of indicators in particular. Meanwhile, the roles of cadastre in land governance continue to be debated in contemporary literature. Some of the key arguments are now discussed.

The debated role of the cadastre in urban land governance can be examined through land administration and urban land governance discourses. Regarding the supportive role, cadastres can support both good and good-enough urban land governance if adequately designed and realized (Table 4.1). Conversely, the cadastre can play an undermining role: Deininger and Feder (2009) reveal that governments in developing countries may take the envisioned benefits of cadastres for granted: the undermining roles are not always considered. Indeed, cadastres may not always serve the society and can even undermine its functioning (Zevenbergen 1999) (Table 4.2).

In summary, it can be observed that urban land governance is the contemporary paradigm through which urbanization and urban environments are studied holistically. Moreover, instances of urban land governance can be understood as residing on a continuum, whereby specific indicators could be used to define cases of good, good-enough, and bad governance, and perhaps even more refined categories. Cadastres are argued to play a determinant role in which a given context sits on the continuum; however, until now most discussions on the role of cadastres concentrate on either of the two extremes. Attention is now given to the methodology used to further develop and synthesize the ideas from this theoretical backdrop.

Methodology

Overall, a literature review focusing broadly on urban cadastres and land governance and more specifically on exemplary cases from the Netherlands and Ethiopia provided the empirical basis. In addition, a systems design approach was used to develop the conceptual model. Details of the specific steps and subsequent integration are now provided.

Regarding the literature review, scientific and gray literature was reviewed. The objective was to draw out conceptual understandings on land governance, and to elucidate how urban cadastres ought to, and ought not to, support urban land governance. In addition, the aim was to scrutinize important exemplary cases linking cadastres and land governance from the global south (Ethiopia) and from the global north (Netherlands). On the last

TABLE 4.1

The Supportive Roles of Cadastres in Urban Land Governance

Supportive Roles of Urban Cadastres	Description of the Different Supporting Roles of Cadastres	Sources
Improves tenure security	Cadastres facilitate creation of land rights and tenure security. Cadastral development encourages informal settlement formalization: it serves to improve tenure security for the unsecured urban poor.	Deininger and Feder (2009), Henssen (2010), van der Molen (2011)
Improves transparency and participation	Cadastral developments, including needs assessments, cadastral policy making, and cadastral surveying and adjudication activities, require participation of citizens and stakeholders: if the cadastre is developed under such conditions, then it enhances transparency.	Roberge et al. (2011), UN-ECE (2005), Williamson et al. (2010)
Provides easy access to information	The digital nature of modern cadastral data, and contemporary information technologies, provides the opportunity for easy access to parcel information.	Arko-Adjei et al. (2010), Deininger and Feder (2009), UN-ECE (2005), Williamson et al. (2010)
Improves governments' and citizens' decision making and efficiency	Easy and timely access to cadastral information reduces land transaction costs and also facilitates decision making and service delivery to citizens. Easy access to information may help citizens to gain access credits, and to decide where to invest. Easy access to cadastral data helps local governments to establish a transparent, equitable, and fair system of land allocation and land taxation.	Deininger and Feder (2009), Henssen (2010), UN-ECE (2005), van der Molen (2011), Williamson et al. (2010)
Reduces rent seeking	Recording parcel geometries and other details including rights, uses, and values reduces the room for rent seeking.	Dale and McLaughlin (1988), van der Molen (2011), Williamson et al. (2010)
Improves equity	Cadastres help to provide an overview of the distribution of land: it identifies who holds what and assists reforms for equitable redistribution. Cadastres support formalization of land rights: the urban poor can get tenure security, and cadastres help to create a sense of equity.	Barry and Fourie (2002), Dale and McLaughlin (1988), UN-ECE (2005)

point, since the 1990s benchmarking and comparative approaches have been frequently applied in land administration and cadastral system comparison and evaluation (Steudler et al. 2004). They provide a means for evaluation of the relative strengths and weaknesses of a specific country's land administration and land governance systems. This type of analysis helps to identify

TABLE 4.2

The Undermining Roles of Cadastres in Land Governance

Undermining Roles of Urban Cadastres	Description of the Different Undermining Roles of Cadastres	Sources
Limits the traditional freedom of use of land and can create mistrust	For customary societies, the cadastre can be regarded as an increasing form governmental control and interference: a loss of freedom and rights is perceived, and mistrust can result.	Arko-Adjei et al. (2010), Deininger and Feder (2009), Dorosh and Thurlow (2011), Henssen (2010)
Can have haphazard consequences for social equity	Cadastres in developing countries are inappropriately used to legalize existing inequity rather than improve it. Tenure security stimulates abnormal increases in land prices, causing socially undesirable land sales that lead to land monopolies, landlessness, disputes, inequity, and social exclusion.	Deininger and Feder (2009), Demsetz (1967), Henssen (2010), van Gelder (2010)
Opens opportunities for rent seeking and litigation	The open and complicated nature of cadastral design and implementation is open to manipulation by influential elites and those who know the system.	Benjaminsen et al. (2008), Loehr (2012), Zevenbergen et al. (2013)
Weakens efficiency of institutions and organizations	Due to its exposure to rent seeking, human resource building can also be hampered. This leads to inefficiency of attempting to implement policy goals.	Deininger and Feder (2009), Payne (2000)
Financially encumbers the state in the short and even longer run	Costs of cadastral development and maintenance are high, while their outcomes may be either contrary to expectation or be manifested in the long run: these combined create financial strains on the national economy.	Atwood (1990), Barry and Fourie (2002), Henssen (2010)

the requirements for systems reengineering and reforms for improvement. Apart from the review of different secondary literature sources, prior knowledge on both the Netherlands and the Ethiopian cadastral and land governance systems by the authors was also taken into account to decide on the cases for comparison.

With respect to the Dutch component of the literature review, the study was limited to English documents (e.g., Wakker et al. 2003; Williamson et al. 2010; Zevenbergen 2002) and focused on the following: governance in general (Huther and Shah 1999; Kaufmann et al. 2010), land policy (Buitelaar 2010; Needham 1997), land laws and legislation (van Rij and Altes 2010), and spatial planning and mapping (van Rij and Altes 2010). Apart from these, land use data from Enschede, Netherlands, were accessed from the ITC Remote Sensing Laboratory, University of Twente, Enschede, for illustrative purposes.

For the Ethiopian exemplary case, results from previous works on Ethiopian urban cadastres and urban land governance were utilized. These had been conducted on multiple case study cities by the authors previously. Specific foci included urban cadastres in Ethiopia (Alemie et al. 2015a), urban land governance in Ethiopia (Alemie et al. 2015c), and spatial analysis of output indicators of land governance such as the cases of informal settlements (Alemie et al. 2015b). In addition, the urban land leasehold laws including proclamation 721/2011 (FDRE 2011a), proclamation 272/2002 (FDRE 2002), and the urban land management policy (FDRE 2011b) were also important sources of information regarding the laws and policies on urban land in Ethiopia.

Following the literature review, a systems approach was applied to develop the conceptual model. A systems approach was utilized because urban environments and the decision-making processes that relate to them can be considered as aggregations of a number of interconnected interactions between different components (Duit et al. 2010). For example, urban land users, urban land, urban land polices and laws, and urban actors contribute their own influence on the urbanization process. Thus, to apply governance concepts to deal with all the issues of urbanization in general and urban people to urban land relationships in particular, an integrated understanding of these components is essential: a systems approach is frequently applied (Checkland 1999). According to Zevenbergen (2002), systems are of collections of elements and associated relationships that combine to explain and deliver a defined goal, such as a land administration system. The systems approach has frequently been applied in land administration, cadastre, land registration, and land tenure analyses (e.g., Rakai 2005; Simbizi et al. 2014; Zevenbergen 2002).

In the context of this work, as already discussed, the urban cadastre is considered as a system that is described by the combination of urban land (subject), urban people (object), and the rights (Henssen 2010; Lemmen 2012; Navratil and Frank 2004). In a formal situation, urban people use urban land through the established legal right. People to land relationships outside or without a legal right are often considered as "informal," at least to the state. In other words, the right is a connector between the subject and the object. The same is also applied in this work. The broader Land Administration Domain Model is built based on these components and their relationships (Lemmen 2012). Similarly, land governance can also be viewed as a system that comprises inputs (policies and laws), processes, information (in this case the urban cadastre), and urban lands. Therefore, the systems approach was employed to integrate the different components of governance, with the resulting indicators, and their spatial outputs to derive a holistic understanding of the types of urban land governance, and the role of urban cadastres, in the urban land governance continuum.

Results

This section presents the literature review results on exemplary cases from the Netherlands and Ethiopia and the design of the conceptual model.

Regarding the results from the literature review of the Netherlands, the country was considered as an exemplary case of good urban governance, supported by a cadastre. The Netherlands' cadastre was described by Williamson et al. (2010) as one of the most successful cadastres globally in terms of its efficient support of the land market, and support to spatial planning and land development activities. Zevenbergen (2002) also revealed that all lands in the Netherlands are recorded under a single system of land registration and this greatly helped to establish an active land market and acceptable levels of legal security. From the technical point of view, Zevenbergen also explained that the national geodetic framework is well established and maintained and the cadastral register is fully digital and can be accessed online. Zevenbergen also described the Dutch cadastre as being legally simple and organizationally sound. The governance index reported by the World Bank (Kaufmann et al. 2010) indicates that the Netherlands scores a high value on all six indicators of governance, including rule of law, government efficiency, quality of regulations, control of corruption, voice and accountability, and political stability. Previous works of the World Bank by Huther and Shah (1999) also revealed that the Netherlands fulfilled the good governance criteria. The Netherlands experience on harmonizing spatial planning with land development is considered an example of successful practice by van Rij and Altes (2010). Rural and urban land accessibility in the Netherlands shows that the existing systems of cadastre, spatial planning, and legal enforcement are interacting harmoniously, at least to a degree. Accordingly, land is offered in good time, for an appropriate function, with a reasonable price (Needham 1992), and a suitable level of scrutiny, and the existence of informal settlements is thus unlikely. Policy and law making in the Netherlands allows the participation of diverse actors including the public, stakeholders, and local governments, among others: the local governments or municipalities are empowered to lead all activities at the operational level including land development (Buitelaar 2010; Needham 1997). Dutch legislation stipulates clear and detailed procedures for actors to follow, for example, the way in which municipalities ought to deal with people to land relationships (van Rij and Altes 2010).

In the case of the Netherlands, the combination of a well-functioning cadastre, spatial planning systems, involvement of diverse actors in policy formulation and implementation, and the significant role of local governments all enhance urban land governance. The Netherlands can be considered an example of good land governance. In other words, the system supports the realization of good governance indicators such as tenure security, a formal land market, equity, information access, transparency, and others.

Regarding the results from the literature review of Ethiopia, previous works by the authors on urban land governance (Alemie et al. 2015c), urban cadastres (Alemie et al. 2015a), and spatial output of governance indicators (Alemie et al. 2015b) provide information regarding its position on the urban land governance continuum. In the three Ethiopian cities studied, it was revealed that tenure insecurity, inequity, and inefficiency of the local government, among others, were major problems in urban land governance in these cities: there was no underlying policy for urban land before 2011; policy and law making and implementation lacked transparency and participation of diverse actors; local governments were weak in terms of human capacity, decision making, and finance, and this led to growth in the number of informal land to people relationships; and the urban cadastres in the three cities showed inefficiency—issues of cost recovery were distant goals. In technical terms, the cadastral systems being created were heterogeneous in terms of the spatial frameworks, software, and surveying instruments used. The spatial or urban land use plans did not cope with rapid urbanization: in most cases, they were obsolete and incomplete. Spatial analyses conducted to map urban land governance indicators, such as informal settlements, depict that informal settlements were increasing in both space and time. However, after the implementation of the 2011 urban land management policy, there was a slight improvement in transparency and also a reduction of informal settlements. Contemporarily, a legal cadastre is considered to be a fit-for-purpose cadastre for the current socioeconomic situation of the country.

Overall, the Ethiopian urban cadastre and urban land governance reflects, at least prior to 2011, the characteristics of somewhat bad urban land governance, albeit with the tendency of moving toward good-enough urban land governance after the 2011 policy implementation.

Regarding the results of the *conceptual modeling*, previous discussions made it clear that urban land governance can be conceptualized through a combined consideration of the inputs, processes, urban cadastres, output indicators, and spatial outputs involved (Figure 4.1). A close study of each component, at both the conceptual and the operational levels, provides insights into the relationships that exist among these components. The overall situation can then be modeled. Each component and the interactions within the conceptual model are now summarized.

Inputs include country-specific urban land policies, laws, and regulations. Inputs are sets of rules that are applied to govern or manage the urban people to urban land relationships in a nation or region. In this sense, inputs are foundational for the urban land governance processes: the way they are formulated and implemented by the actions of actors are central in understanding governance. Overall, the contribution of policies and laws to urban land governance, or sustainable development in general, depends on the nature and capabilities of actors in both formulating and implementing the inputs.

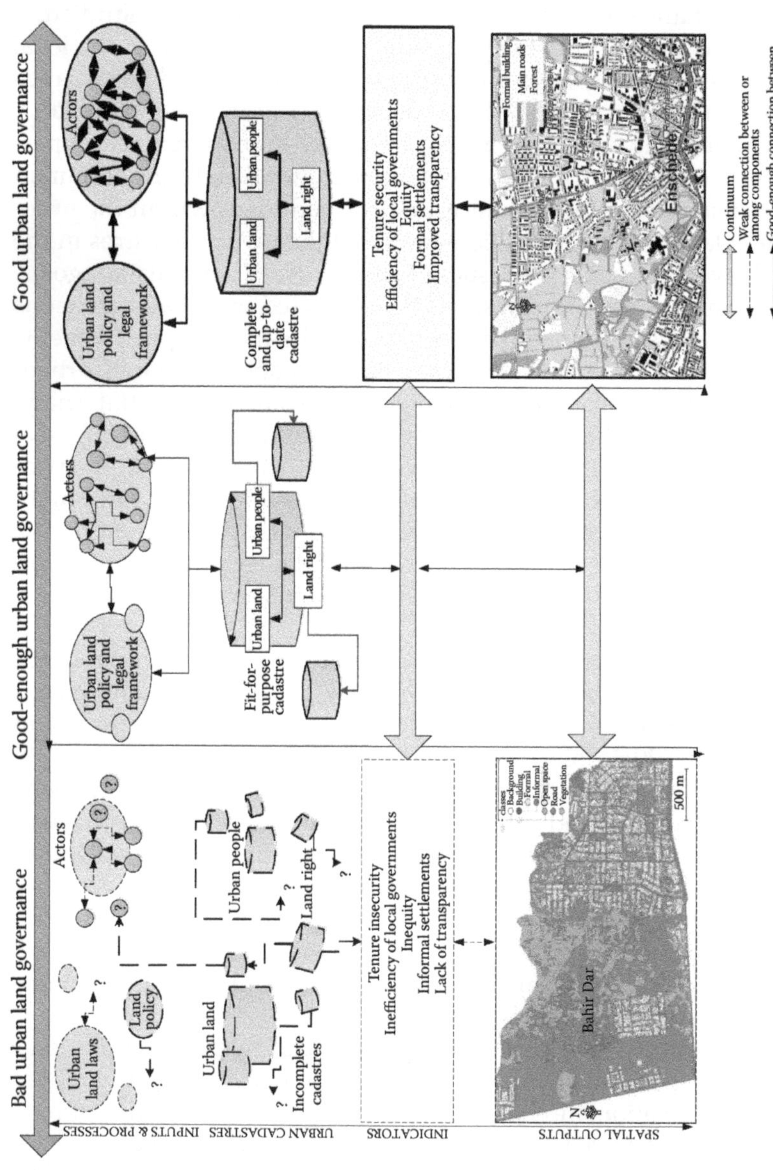

FIGURE 4.1
A conceptual and analytical framework depicting the types of urban land governance across a land governance continuum.

Processes refer to the interaction between different actors including urban people, who are the primary user of the urban land, and the different stakeholders that work in the urban land sector, such as municipalities. Achievement of a policy goal is a function of the roles and capacities of actors, in both policy making and implementation. Actors transform the inputs that exist on paper into physical activities on urban lands. This implies that actors are bridges to connect the inputs with the outputs on urban land. Overall, processes of policy making and implementation by actors contribute to the success, or failure, of urban land governance and urban cadastres.

Urban cadastres refer to the spatial aspect and attribute information about urban land (subject), urban people (object), and the relationships created through legal rights. This information is useful in decision making and governance of urban land. The qualities of decision making, in this aspect, rely on the availability and quality of the information being used (Dale and McLaughlin 1988). Overall, cadastral information is seen as an essential input to both formulate and implement policies and laws related to urban land.

Indicators of urban land governance are derivatives of the interaction between the previously discussed components; but, they can also be considered entities in their own right. For example, if the urban land policy is formulated based on the involvement of actors in a transparent way, and decision making is supported by appropriate information sources (e.g., cadastral information), then the outputs of decision making and governance can benefit the majority of the urban people. From the literature review, results on the practices of Ethiopia, together with the contemporary government's focus in the urban land sector, the following indicators were deemed most important and were used in the conceptual model. These include evidence of tenure security, equity, local government capacity, informal settlements, and transparency.

Spatial outputs are results of urban land governance processes and manifested on urban lands. They could also be considered a form of indicator; however, in the conceptual model presented they are raised to a higher status for interpretive ease. Spatial analysis of satellite or aerial images on the urban environment can provide a reasonably objective status on urban land governance. The spatial and temporal characteristics of urban lands differ from one type of urban land governance to another, across the governance continuum. As an example, informal settlements can be identified spatially: they have specific characteristics that can be captured by remote sensing techniques. Informal settlements are also known to be the combined result of tenure insecurity, inequity, weak local government, and lack of transparency. However, the latter elements are often difficult to expose, or measure: they are subjective. Thus, the spatial outputs of urban land governance indicators are seen as highly tangible and highly objective constructs: they complement the other components of the model and close the link between inputs, processes, urban cadastres, indicators, and outputs.

Discussion

The systems approach to developing conceptual models can be applied in two ways. The first approach relies only on the use of literature sources to understand and clarify theoretical concepts and their conceptual relationships. The assumption is that there will not be uncertainties during the analysis process. A conceptual model design process follows (Simbizi et al. 2014; White et al. 2009). Testing of the model is conducted thereafter. The second approach prefers to first draw a clear understanding of both the conceptual and the operational levels through case studies or exemplary cases. The information from these will be input for the design of the conceptual model (Faehnle and Tyrväinen 2013). In this chapter, the second approach is applied. This is because there are uncertain situations that hinder proper implementation of the model that is developed solely from the consideration of theoretical concepts. Thus, such conceptual models cannot always be useful when attempting to solve real urban land governance problems. Especially in cadastre and land governance discourses, where the concepts are subjected to both frequently evolving theories and country-specific policies, laws, and actors, a combination of the two approaches sharpens the conceptual models. As discussed, the operational situations of cadastre and land governance of the Netherlands and Ethiopia informed the modeling. These were obtained from the review of literature that was originally the result of detailed empirical and policy analysis.

The conceptual model in Figure 4.1 illustrates the types of land governance in a continuum, accompanied by the different components including input, process, cadastre, indicators, and spatial outputs. The inputs and processes influence the cadastre, the resulting land governance indicators, and their spatial outputs. This suggests important messages for decision makers, especially in those countries where the benefits of cadastres are taken for granted: both success of cadastres and achieving improved land governance indicators, in any country, are dependent on the inputs and processes. In other words, improvement of the inputs and processes needs to be the first step toward improving the cadastre and land governance more broadly.

The conceptual model also shows the nature of the actors in the three urban land governance types: they formulate and implement policies and laws. For example, in a bad governance situation actors are few and their communication is restricted: there is no integration, decision making is unidirectional, and it is often a top–down process. In this case, either policies may not exist (e.g., the cases of Ethiopia before 2011) or if policies exist they are poorly formulated and implemented due to the poor performance of actors at lower levels, malicious or otherwise. Under these contexts, land rights may not exist or may be poorly defined, and consequently tenure insecurity problems cannot be resolved. This also has effects on the recordation of land to people relationship information in a cadastre: the cadastre cannot support the improvement of urban land governance. Indeed, it may play an undermining role as seen in Ethiopia prior to 2011 (Alemie et al. 2015a).

In a good governance situation, diverse actors are involved in an integrated way, during both policy making and its implementation. The policy and laws will clearly address how to deal with people to land relationships, such as urban land rights, and guarantee tenure security. This situation creates an efficient urban cadastre that can support a great deal of planning and land management, as depicted in the Dutch examples: urban lands are utilized based on the legal rights recorded in the urban cadastre and designated uses in the land use plan. Good quality output indicators, for example, "no informal settlements," are achieved (see the map of part of Enschede at the bottom right of Figure 4.1). Overall, cadastres significantly contribute to the improvement of urban land governance.

Attempting good governance, particularly in countries with institutional and organizational deficiencies, however, is not an easy task and cannot always work: it remains a distant goal. The capacity of actors, simplicity of policies and laws, and efficiency of cadastres are all products of the context. In this respect, considering an intermediate option, such as good-enough urban land governance, provides another option. Figure 4.1 shows that the number of actors in the good-enough governance case is higher than that in bad governance and lower than that in a good governance situation. In addition, policies and laws exist that are in line with governance concepts. In this respect, the situation in Ethiopia post 2011 can be an example. The first urban land management policy was established, laws to improve informal settlements such as formalization laws were issued, and actors tended to commence exercising a mix of bottom–up and top–down decision-making activities during policy implementation. The government identified issues of tenure insecurity and informal settlements as the major problems in contemporary urban development. Following this, the development of an urban legal cadastre was identified as fit-for-purpose in the immediate context: it would help to solve immediate problems already identified, as opposed to the previous ambitious multipurpose cadastre. According to Alemie et al. (2015c), some improvements were evident at the initial stages of the 2011 policy implementation. These included improvements in transparency and reductions in rent seeking. In this respect, there appears to be a gradual shift toward good-enough urban land governance in the context of the three Ethiopian cities studied.

The conceptual model shows that the urban land governance situation of any country, and its cities, can fall in any of the three types of urban land governance in the continuum: the model applies beyond the contexts of the Netherlands and Ethiopia. In addition, this work adds further conceptual scrutiny to the existing land governance works of international organizations such as the World Bank and the Food and Agriculture Organization (FAO 2012). For example, the "Voluntary Guidelines" are considered quite comprehensive; however, attempting to satisfy all the land governance principles discussed in the guidelines at once, or in a short time frame, seems to be unrealistic, at least in the current status quo of most developing countries. In many cases, a focus on good-enough land governance appears to be a more workable option.

The contemporary situations in Ethiopia show a move toward fit-for-purpose land administration and good-enough land governance. In line with the contemporary socioeconomic development in Ethiopia, a legal cadastre is found to be fit-for-purpose and relevant to support the marketization of urban land. The tendency in Ethiopia toward a fit-for-purpose cadastre and good-enough land governance can be a lesson for other countries with similar socioeconomic situations and current land governance problems.

Conclusions

The conceptual model developed in this chapter brings a holistic understanding of urban land governance and its relationship with cadastre. It differs from other existing models in several aspects: (1) it presents a continuum of urban land governance and cadastral interactions—it presents more than a simple positive or negative relationship; (2) it is based on both theoretical concepts and empirical evidence, whereas existing models are usually confined to theoretical concepts; (3) the model considers inputs, processes, and outputs of urban land governance, whereas existing models are inclined to focus only on inputs and processes; and (4) the model considers both the social and spatial dimensions of urban land governance, as opposed to existing models focused only on social understandings. The cases of specific countries are provided for the three types of urban land governance: bad urban land governance is illustrated by the situation in Ethiopia before 2011, the tendency toward good-enough urban land governance is exhibited in contemporary Ethiopia, and good urban land governance situation is demonstrated in the cases of the Netherlands. It is further implied that the model can be applied to different country contexts. The model also shows that the roles of cadastre in urban land governance range from minimal to substantial. For example, fit-for-purpose cadastres can support the realization of good-enough urban land governance. Future works should focus on scrutinizing the components of the conceptual model, examination of the fit-for-purpose cadastre (the legal cadastre in the contemporary situation in Ethiopia), and its contribution to good-enough urban land governance.

References

Africa Union Commission 2010. Land Policy in Africa: A Framework to Strengthen Land Rights, Enhance Productivity and Secure Livelihoods. Addis Ababa, Ethiopia, African Union Commission.

Alemie, B. K., Bennett, R. M. and Zevenbergen, J. 2015a. Evolving urban cadastres in Ethiopia: the impacts on urban land governance. *Land Use Policy*, 42, 695–705.

Alemie, B. K., Bennett, R. M. and Zevenbergen, J. 2015b. A socio-spatial methodology for evaluating urban land governance: the case of informal settlements. *Journal of Spatial Science*, doi: http://dx.doi.org/10.1080/14498596.2015.1004654. [In Press].

Alemie, B. K., Zevenbergen, J. and Bennett, R. 2015c. Integrating the layers: an analysis of urban land governance in contemporary Ethiopia. *Administration and Society*. [Under Review].

Arko-Adjei, A., Jong, J. D., Zevenbergen, J. and Tuladhar, A. 2010. "Customary Tenure Institutions and Good Governance." FIG Conference on Facing the Challenges—Building the Capacity, Sydney, Australia, FIG.

Arnold, C. A. 2002. The reconstitution of property: property as a web of interests. *Harvard Environmental Law Review*, 26, 281–364.

Atwood, D. A. 1990. Land registration in Africa: the impact on agricultural production. *World Development*, 18(5), 659–671.

Barry, M. and Fourie, C. 2002. Analysing cadastral systems in uncertain situation: a conceptual framework based on soft system theory. *International Journal of Geographical Information Science*, 16(1), 23–40.

Benjaminsen, T. A., Holden, S., Lund, C. and Sjaastad, E. 2008. Formalisation of land rights: some empirical evidence from Mali, Niger and South Africa. *Land Use Policy*, 26, 28–35.

Bennett, R. and Alemie, B. K. 2015. Fit-for-purpose land administration: cases from urban and rural Ethiopia. *Survey Review*. [In Press].

Bennett, R., Rajabifard, A., Williamson, I. and Wallace, J. 2012. On the need of national land administration infrastructure. *Land Use Policy*, 29(1), 208–219.

Benda-Beckmann, F. V., Benda-Beckmann, K. V. and Wiber, M. 2006. *Changing Properties of Property*, New York, Berghahn Books.

Bogaerts, T. and Zevenbergen, J. 2001. Cadastral systems—alternatives. *Computers, Environment and Urban Systems*, 25, 325–337.

Buitelaar, E. 2010. Cracks in the myth: challenges to land policy in the Netherlands. *Tijdschrift Voor Economische En Sociale Geografie*, 101(3), 349–356.

Burns, T. and Dalrymple, K. 2008. "Conceptual Framework for Governance in Land Administration." FIG Working Week, Stockholm, Sweden, FIG.

Checkland, P. 1999. *Soft Systems Methodology: A 30-Year Retrospective*. Chichester, England, John Wiley.

Dale, P. F. and McLaughlin, J. D. 1988. *Land Information Management; An Introduction with Special Reference to Cadastral Problems in Third World Countries*. New York, Oxford University Press.

Dawson, R. J., Flacke, J., Salvia, M. and Pietrapertosa, F. 2014. Understanding cities: the imperative for integration. In: Dawson, R. J., Wyckmans, A., Heidrich, O., Köhler, J., Dobson, S. and Feliu, E. (eds.). *Understanding Cities: Advances in Integrated Assessment of Urban Sustainability*. Tyne, United Kingdom, Newcastle University: Centre for Earth Systems Engineering Research (CESER).

de Soto, H. 2000. *The Mystery of Capital: Why Capitalism Triumphs in the West and Fails Everywhere Else*. London, United Kingdom, Bantam press/Random House.

Deininger, K. and Feder, G. 2009. Land registration, governance, and development: evidence and implications for policy. *World Bank Research Observer*, 24(2), 233–266.

Demsetz, H. 1967. Toward a theory of property rights. *The American Economic Review*, 57(2), 347–359.

Dobson, S., Hurtado, S. D., Coelho, D., Wyckmans, A. and Schmeidler, K. 2014. A policy and governance context for integrated urban sustainability strategies. In: Dawson, R. J., Wyckmans, A., Heidrich, O., Köhler, J., Dobson, S. and Feliu, E. (eds.). *Understanding Cities: Advances in Integrated Assessment of Urban Sustainability*. Tyne, United Kingdom, Newcastle University: Centre for Earth Systems Engineering Research (CESER).

Dorosh, P. and Thurlow, J. 2011. "Urbanization and Economic Transformation: A CGE Analysis for Ethiopia." Ethiopia Strategy Support Program II (ESSP II) Working Paper 14. Addis Ababa, Ethiopia, Development Strategy and Governance Division, International Food Policy Research Institute.

Duit, A., Galaz, V., Eckerberg, K. and Ebbesson, J. 2010. Governance, complexity, and resilience introduction. *Global Environmental Change-Human and Policy Dimensions*, 20(3), 363–368.

Enemark, S. 2005. Understanding the Land Management Paradigm. Innovative Technologies for Land Administration, Madison, WI, FIG.

Faehnle, M. and Tyrvainen, L. 2013. A framework for evaluating and designing collaborative planning. *Land Use Policy*, 34, 332–341.

FAO 2007. Good Governance in Land Tenure and Administration. Rome, Italy, FAO.

FAO 2012. Voluntary Guidelines on the Responsible Governance of Tenure of Land, Fisheries and Forests in the Context of National Food Security. Rome, Italy, FAO.

FDRE 2002. A Proclamation to Provide for the Re-Enactment of Lease Holding of Urban Lands. Proclamation No. 272/2002. Addis Ababa, Ethiopia, Negarit Gazeta.

FDRE 2011a. A Proclamation to provide for Lease Holding of Urban Lands. Proclamation No. 721/2011. Addis Ababa, Ethiopia, Negarit Gazeta.

FDRE 2011b. Urban Land Management Policy. Addis Ababa, Ethiopia, Ministry of Urban Development and Construction.

Fekade, W. 2000. Deficits of formal urban land management and informal responses under rapid urban growth, an international perspective. *Habitat International*, 24(2), 127–150.

FIG 2014. "Fit-for-Purpose Land Administration." FIG Publication 60. Copenhagen, Denmark, FIG.

Foster, K. A. 2000. "Smart Governance, Smart Growth." Lincoln Institute of Land Policy conference paper. Ithaca, NY, Lincoln Institute of Land Policy.

Gao, B. Y., Liu, W. D. and Dunford, M. 2014. State land policy, land markets and geographies of manufacturing: the case of Beijing, China. *Land Use Policy*, 36, 1–12.

GLTN 2008. Secure Land Rights for All. Nairobi, Kenya, GLTN/UN-Habitat.

Grindle, M. S. 2011. Good enough governance revisited. *Development Policy Review*, 29, 199–221.

Guo, R., Li, L., Ying, S., Luo, P., He, B. and Jiang, R. 2013. Developing a 3D cadastre for the administration of urban land use: a case study of Shenzhen, China. *Computers, Environment and Urban Systems*, 30, 46–65.

Henssen, J. 2010. "Land Registration and Cadastre Systems: Principles and Related Issues." Lecture notes, masters program in land management and land tenure. Munchen, Germany, TU.

Huther, J. and Shah, A. 1999. *Applying a Simple Measure of Good Governance to the Debate on Fiscal Decentralization*. Washington DC, World Bank.

Kaufmann, D., Kraay, A. and Mastruzzi, M. 2010. "The Worldwide Governance Indicators: A Summary of Methodology, Data and Analytical Issues." World Bank Policy Research Working Paper No. 5430. Washington DC, World Bank.

Lemmen, C. 2012. "A Domain Model for Land Administration." PhD diss., Delft, The Netherlands, Netherlands Geodetic Commission.

Ligtenberg, A., Beulens, A., Kettenis, D., Bregt, A. K. and Wachowicz, M. 2009. Simulating knowledge sharing in spatial planning: an agent-based approach. *Environment and Planning B-Planning and Design*, 36(4), 644–663.

Loehr, D. 2012. Capitalization by formalization?—challenging the current paradigm of land reforms. *Land Use Policy*, 29, 837–845.

McNeill, D., Bursztyn, M., Novira, N., Purushothaman, S., Verburg, R. and Rodrigues, S. 2014. Taking account of governance: the challenge for land-use planning models. *Land Use Policy*, 37, 6–13.

Melesse, M. 2005. "City Expansion, Squatter Settlements and Policy Implications in Addis Ababa: The Case of Kolfe Keranio Sub-City." Working papers on population and land use change in central Ethiopia, No. 2. Trondheim, Norway. NTNU and Addis Ababa University.

Navratil, G. and Frank, A.U. 2004. Processes in a cadastre. *Computers, Environment and Urban Systems*, 28, 471–486.

Needham, B. 1992. A theory of land prices when land is supplied publicly—the case of the Netherlands. *Urban Studies*, 29(5), 669–686.

Needham, B. 1997. Land policy in the Netherlands. *Tijdschrift Voor Economische En Sociale Geografie*, 88(3), 291–296.

Obeng-Odoom, F. 2012. Land reforms in Africa: theory, practice, and outcome. *Habitat International*, 36(1), 161–170.

Palmer, D., Fricska, S. and Wehrmann, B. 2009. Towards Improved Land Governance. Rome, Italy, FAO/UN-Habitat.

Payne, G. 2000. "Urban Land Tenure Policy Options: Titles or Rights?" Paper presented at the World Bank Urban Forum. Westfields Marriott, Virginia.

Payne, G. 2005. Getting ahead of the game: A twin-track approach to improving existing slums and reducing the need for future slums. *Environment and Urbanization*, 17(1), 135–145.

Rakai, M. E. T. 2005. "A Neutral Framework for Modeling and Analysing Aboriginal Land Tenure Systems." PhD diss., University of New Brunswick, Fredericton, Canada.

Rakodi, C. 2003. Politics and performance: the implications of emerging governance arrangements for urban management approaches and information systems. *Habitat International*, 27(4), 523–547.

Roberge, A., Bouthillier, L. and Mercier, J. 2011. The gap between theory and reality of governance: the case of forest certification in Quebec (Canada). *Society and Natural Resources: An International Journal*, 24, 656–671.

Silva, M. A. 2005. "Modelling Causes of Cadastral Development—Cases in Portugal and Spain During the Last Two Decades." PhD diss., Aalborg University, Aalborg, Denmark.

Simbizi, M. C. D., Bennett, R. M. and Zevenbergen, J. 2014. Land tenure security: revisiting and refining the concept for Sub-Saharan Africa's rural poor. *Land Use Policy*, 36, 231–238.

Steudler, D., Rajabifard, A. and Williamson, I. P. 2004. Evaluation of land administration systems. *Land Use Policy*, 21(4), 371–380.

Szeftel, M. 2000. Between governance and under-development: accumulation and Africa's 'catastrophic corruption'. *Review of African Political Economy*, 27(84), 287–306.

Thuo, A. D. M. 2013. Impacts of urbanization on land use planning, livelihood and environment in the Nairobi rural-urban Fringe, Kenya. *International Journal of Scientific and Technology Research*, 2(7), 70–79.

UN-ECE 2005. Land Administration in the UNECE Regions—Development Trends and Main Principles. New York, Geneva, Switzerland, ECE.

UN-FIG 1999. The Bathurst Declaration on Land Administration for Sustainable Development. Bathurst, Australia, FIG.

UN-Habitat 2012. "State of the World's Cities Report 2012/2013: Prosperity of Cities." Nairobi, Kenya, UN-Habitat.

van der Molen, P. 2011. Land administration. In: Lemmens, M. (ed.). *Geo-Information, Geotechnologies and the Environment 5*. New York, Springer Science + Business Media B.V.

van Gelder, J. -L. 2010. What tenure security? The case for a tripartite view. *Land Use Policy*, 27, 449–456.

van Rij, E. and Altes, W. K. 2010. Looking for the optimum relationship between spatial planning and land development. *Town Planning Review*, 81(3), 283–306.

Wakker, W. J., van der Molen, P. and Lemmen, C. 2003. Land registration and cadastre in the Netherlands, and the role of cadastral boundaries: the application of GPS technology in the survey of cadastral boundaries. *Journal of Geospatial Engineering*, 5(1), 3–10.

White, R. M., Fischer, A., Marshall, K., Travis, J. M. J., Webb, T. J., di Falco, S. et al. 2009. Developing an integrated conceptual framework to understand biodiversity conflicts. *Land Use Policy*, 26(2), 242–253.

Williamson, I., Enemark, S., Wallace, J. and Rajabifard, A. 2010. *Land Administration for Sustainable Development*. Redlands, California, Esri Press.

Zevenbergen, J. 1999. Are cadastres really serving the landowner? In: *Proceedings of 21st Urban Data Management Symposium: Information Technology in the Service of Local Government Planning and Management*, Venice, Italy, April 21–23, 1999, UDMS, Delft, The Netherlands.

Zevenbergen, J. 2002. "Systems of Land Registration—Aspects and Effects." PhD diss., Delft University of Technology, Delft, The Netherlands.

Zevenbergen, J., Augustinus, C., Antonio, D. and Bennett, R. 2013. Pro-poor land administration: principles for recording the land rights of the underrepresented. *Land Use Policy*, 31, 595–604.

5

Land Administration for Post-Conflict Contexts

Dimo Todorovski, Jaap Zevenbergen, and Paul van der Molen

CONTENTS

Introduction

Conflict, land, and property meet each other at every point in the conflict cycle, as experienced in many countries (Guterras 2009). A critical gap relating to the land component was identified in 2005 (OCHA 2005). Awareness on the importance of addressing housing, land, and property (HLP) in post-conflict contexts has increased since then, in both literature and practice. Despite this heightened awareness, knowledge of the nature of these problems and potential ways to address them is limited when it comes to state building (Kato 2014). More specifically, a knowledge gap identified in the theory relates to the potential that land administration has within the post-conflict state building process (Todorovski et al. 2012).

New thinking on post-conflict state building argues that land administration has a particular role to play. Both practice and literature show that HLP and land administration are always negatively affected by conflict. Further, if they are not addressed appropriately in a post-conflict context they could spawn new disputes or even armed conflict (Lewis 2004). This might complicate the fragile post-conflict situation and the success of post-conflict state building efforts (Takeuchi 2014). Therefore, post-conflict state building with a specific focus on the role of land needs better understanding.

The aim of this chapter is to recognize the characteristics of post-conflict state building, with a specific focus on HLP and land administration. The section on the theoretical perspective is based on the theory about (1) post-conflict state building and (2) land in post-conflict contexts. The methodology section explains the research gap, presents the case of Rwanda, and describes the method of data collection. The case study of Rwanda provides evidence about developments regarding land in post-conflict contexts. The discussion section identifies the relevant HLP and land administration issues for post-conflict state building. The discussion starts with the background factors of land, conflict, and post-conflict contexts and then continues in a structure of the main characteristics of war-torn societies: institutional weaknesses, social and economic problems, and serious security problems. This chapter finishes by articulating several key summary statements.

One country where the characteristics of post-conflict state building, focusing on HLP and land administration, can be further explored is Rwanda. Rwanda witnessed violent conflicts that resulted in large-scale population displacement and ended with one of the worst genocides known in recorded history. The displaced population started to return after the end of the conflict in 1994 when the security and humanitarian situation on the ground improved (Prunier 1997). In a short period after the end of the conflict, land issues emerged for a large number of people. Tackling HLP and land administration was a very sensitive activity that required specific attention and an approach that corresponded to the local circumstances (Potel 2014). Land-sharing processes, allocation of state land, housing and village settlement for the returnees, and land administration developments are observed in this chapter within the broader post-conflict state building concept.

Theoretical Perspective

This section, based on the available theory, provides an overview of post-conflict state building and land in post-conflict contexts.

First, regarding post-conflict state building, causes of armed conflicts are many and stem from territory, ideology, dynastic legitimacy, religion, language, ethnicity, self-determination, resources, markets, dominance, equality, and of course revenge (Heinz-Jürgen et al. 2006). Wars today kill fewer people than in the past; however, greater numbers of civilians are exposed and vulnerable to violence. The United Nations (UN), reporting on the status of force displacement worldwide, at the end of 2013 estimated 51.2 million displaced persons (UNHCR 2014). According to FAO (2005), the violent conflict is over and the post-conflict period starts when main hostilities have ceased, a peace agreement document (PAD) is signed, and international assistance can be provided. The post-conflict phases should not be understood as absolute, fixed, or time bound or having clear boundaries. However, scholars agree that a PAD is the foundation for the post-conflict state building period that follows.

After a PAD has been signed, the first challenge that post-conflict states face is the process of keeping peace, which is understood as a situation with absence of armed conflict, and at least a modicum of political process. Peace building is described as the whole of actions undertaken by international or national actors to institutionalize peace. Peace building requires some form of "doing justice," and therefore application of the rule of law becomes a necessary perspective for looking at the given post-conflict environment (Call and Cousens 2008).

To develop stability and socioeconomic progress, it is recognized that the rule of law is critical to states emerging from a long period of conflict and misrule. It is important that the rule of law is not only provided for in the law but also practiced by the officials of the state, allows participation of the citizens, and is enforced by courts (Rugege 2013).

State building is defined as follows:

> Purposeful action to build capacity, institutions and legitimacy of the state in relation to an effective political process to negotiate the mutual demands between the state and societal groups (OECD 2008).

Fukuyama defines state building as follows:

> The creation of new government institutions and the strengthening of existing ones (Fukuyama 2004).

In post-conflict contexts, the aim of state building is building effective systems and institutions of government; establishment of trust and mutual accountability; notion of rights and obligations by citizens; and a political agenda aiming at the development of an inclusive state in support of an equitable economic, political, and social order (Brahimi 2007).

Ball (2001), observing further state building in post-conflict contexts, distinguishes three basic characteristics of war-torn societies. The first characteristic comprises institutional weaknesses, like nonparticipatory and malfunctioning political and judicial systems, strong competition for power instead of attention to governing, a limited legitimacy of political leaders, and no consensus on which way society should go. The second characteristic comprises economic and social problems: destroyed or decaying social and economic infrastructure, an increase of the illegal economy, people reverting to subsistence activities, hatred among population groups, and conflicts over land and property. Finally, these societies have to cope with serious security problems: huge quantities of small arms freely circulating among the population, political influence of the armed forces, and demobilization and disarmament (Ball 2001).

In post-conflict situations, international actors with available resources and skills should facilitate local processes and create a space for local actors, who are the main workforce. Together, they should define and consolidate their policies to build responsive, resilient, and robust institutions (Chesterman et al. 2005).

At the end of this chapter, under Discussion, the widely accepted characteristics of war-torn societies, developed by Ball (2001), are used as a primary lens through which HLP and land administration within post-conflict state building are observed.

Second, regarding land in post-conflict contexts, it is observed that although most violent conflicts are not caused by conflicts over land per se, almost every major eruption of violent conflict has a land dimension (Putzel 2009). EU-UN (2012) sees land as politically too sensitive or technically too complicated to be tackled early in the post-conflict period. Land issues always arise during conflict, but they become very complex immediately after the conflict. In a short period of time, a mass population returns to their places of origin and a common challenge for all post-conflict contexts is resettlement of the displaced population in an equitable process. Old properties and houses are usually destroyed, damaged, or illegally occupied by secondary occupants (Leckie 2000). The moment of return of people is critical and a possibility for the eruption of new conflicts. Disputes over land are a very frequent problem in post-conflict settings. Thus, land dispute resolution mechanisms are viewed as a necessary tool that should contribute to the peace process (Unruh 2001).

Hollingsworth (2014) further developed the factors affecting land issues in the post-conflict period. Under "institutional characteristics," the factors affecting land are as follows: international presence, UN, inappropriate legislation, unavailable land administration systems, lack of appropriately trained and skilled staff, and manipulation of land records by politically powerful people. Under "social and economic characteristics," displacement and destruction of houses and properties may be the most serious consequence of armed conflict, together with loss of proof of identity and proof of land rights and ownership. Under "security characteristics," factors are increased risk of secondary conflict in relation to land and conflicts between hosts and the displaced population (Hollingsworth 2014).

Recognizing the important role that land is playing in post-conflict circumstances drives this research to explore the theory on how land is administered in these circumstances. UN/ECE's Land Administration Guidelines refer to land administration as follows:

> The process of determining, recording and disseminating information about tenure, value and use of land when implementing land management policies (UN/ECE 1996).

Land administration is the appropriate instrument for implementing national land policies. It performs a number of functions including supporting the establishment of the land market, land use organization, setting land taxes, and management of state land. The goal of land administration processes is to support the implementation of land policies using the aspects of land management (van der Molen 2002). According to Wehrmann (2005), the transfer of criteria of good governance to post-conflict land policy and land management would likely provide a good basis for sustainable and low-conflict development. The establishment of such a framework is of crucial importance, especially in situations of post-conflict countries (Wehrmann 2005).

Specific land-related challenges that come up in the conflict and post-conflict contexts are loss/destruction of properties, secondary occupations, landlessness, access to land, not functional land administration systems, forced transactions, emergency occupation of land, and HLP rights (UN-HABITAT 2007). The most obvious negative impact for land administration follows from the loss of staff and records. Staff can be killed, (forced to) leave the area, or be not able or willing to return to their jobs within land administration systems (Zevenbergen and Burns 2010). Incomplete, out-of-date, or contested land records can pose a threat to tenure security and the overall peace situation. Issues about land records in post-conflict situations that require appropriate attention are as follows: inadequate land records; fragmented responsibility for land records; lost, stolen, fraudulent, and altered land records; and women and child's property and inheritance rights (UN-HABITAT 2009). HLP and land administration issues are always affected by the conflict, and if not addressed properly they can be the cause of secondary conflicts (Todorovski et al. 2012). Experiences from other post-conflict contexts, show that only in cases of Kosovo and Timor-Leste specific land management and land administration activities were integrated in the PAD and in UN operations (AD 1999; UN 1999).

Methodology

From the theoretical perspective, it can be derived that the characteristics of post-conflict state building that relate to land, and more specifically land administration, need better understanding. The potential for land administration to contribute to the post-conflict state building process constitutes a clear knowledge gap (Figure 5.1).

The knowledge gap requires further study. The phenomenon is identifiable in contemporary practical contexts; however, the boundaries between the phenomenon and the context are not clear. Yin defines a case study as follows:

> An empirical inquiry that investigates a contemporary phenomenon within its real life context, especially when the boundaries between phenomenon and the context are not clearly evident (Yin 2003).

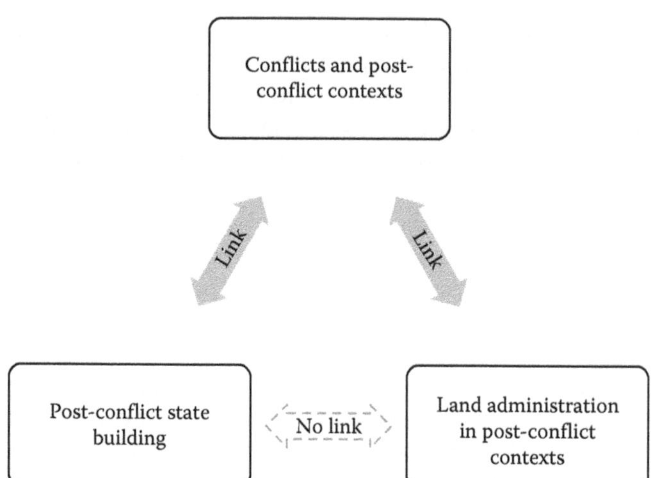

FIGURE 5.1
Framework representing the gap in knowledge. (Adapted from Todorovski, D. et al., "Can Land Administration in Post-Conflict Environment Facilitate the Post-Conflict State Building?—a Research Problem," FIG Working Week: Knowing to Manage the Territory, Protect the Environment, Evaluate the Cultural Heritages, Rome, Italy, 2012.)

Accordingly, a case study was the most suitable approach to be applied to the work.

One of the cases where the gap in knowledge can be further explored is Rwanda. Rwanda is a country that has experienced violent conflicts and adopted developments in the area of post-conflict land administration within a broader post-conflict state building framework. More precisely, Eastern Rwanda is chosen as a case study area: in this area, the majority of the refugees settled after their return. Data collection was executed in three separate fieldwork activities (Manirakiza 2014; Potel 2014; Songo 2014). The selected locations in Eastern Rwanda for data collection were Gasabo district (first fieldwork), Kayonza and Ngoma districts (second fieldwork), and Rukara and Gahini sectors in Kayonza district (third fieldwork).

This study used primary data from the three fieldwork activities, secondary data collected from various organizations dealing with land issues and was supported by a comprehensive literature review. The "Results" section presents the findings from the collected evidence about Rwanda, with respect to (1) land and conflict relation, (2) HLP and land administration in post-conflict contexts, and (3) post-conflict state building (as they occurred during the conflict and the post-conflict period). The section "Discussion" takes a more generalized approach and aims to develop more global understandings. The section starts with the background factors of the land, conflict, and post-conflict contexts. Then, the characteristics of post-conflict state building are elaborated, focusing on land in the structure of the characteristics of war-torn societies (Ball 2001). Based on the discussions, a table of characteristics of post-conflict state building focusing on land is developed.

Results

Land, conflict, and post-conflict contexts: before and during the colonial era, land was used to divide the Rwandan population along ethnic lines. For the conflict in Rwanda, which finished in 1994, land issues were considered as the major cause: they were used as a fuelling factor for the increase in ethnic tensions (Magnarella 2005). Political representation and unresolved governance issues were mentioned as additional causes (Nyamwasa et al. 2010).

Among the characteristics of violent conflict, the most serious concerns were the number of people killed, widespread destruction of properties and infrastructure, and displacement. The first violent conflict in 1959 in Rwanda resulted in half a million refugees crossing the borders of neighboring countries. For the period after this conflict, the government of Rwanda used land as a political tool, redistributing abandoned properties to their political followers and military officers, causing illegal occupation by secondary occupants (Potel 2014). The second violent conflict in Rwanda finished in 1994 with a genocide that caused more than one million people to die, a large number of internally displaced persons (IDPs), and more than two and a half million refugees (Prunier 1997).

A peace treaty was signed in Arusha on July 25, 1993 (PAD 1993). After signing the PAD, the security situation was fragile and tense, with a large military presence, presence of UN troops, and lots of sporadic conflicts: finally, the genocide happened in the first half of 1994. Only after the victory of the Rwanda Patriotic Front (RPF), which took control over the whole country by mid-July 1994, did the security situation improve. Displaced people began returning. In process terms, this moment can be considered as the start of the post-conflict period. Within such a post-conflict period, one can distinguish the emergency phase, early recovery phase, and reconstruction phase (FAO 2005). The emergency phase in post-conflict Rwanda was from 1994 to 1997 and was marked by unity and reconciliation activities. After the humanitarian crisis, the main focus was on the provision of shelter, food, and other basic living conditions for the population returning en masse (Murekezi 2012). Actors on the ground did not seem to understand the importance of land issues at this time: these were totally out of the focus of activities (Bruce 2007). The early recovery phase can be seen as being from 1997 until the end of 2002: this period involves the development of a legal framework, national policy developments, new government structures, and strategies for their implementation. The period after 2003 can be observed as the reconstruction phase: it concerned implementation and execution of the legal frameworks, national policy, and programs (Songo 2014).

As identified in the theory, land starts to come to the forefront specifically during the emergency post-conflict phase, when people en masse return to their place of origin in a short period of time (Pantuliano 2009; Unruh and Williams 2013). In Rwanda, a large number of people displaced during the first conflict returned immediately after the conflict ended in 1994. As per the recommendation in the PAD, the majority of returnees did not go back to their places of origin; instead, they were allocated with the available state land, for example, land in the national parks for housing and village settlements. The rest of the returnees went back and repossessed their old properties. The return continued until the end of 1996, and the last wave of one and a half million people who returned were displaced persons from 1994. At the end of 1996, Rwanda witnessed a situation with almost two and a half million returnees who needed shelter. Land emerged as a very sensitive issue, which required a specific approach that corresponded to the local circumstances (Potel 2014).

HLP and land administration in post-conflict contexts: different solutions to land-related problems occur in different post-conflict cases. Emergency measures such as the creation of IDP camps and other settlements, on state public or private land, are de facto interventions in land management and land tenure (de Waal 2009). The Arusha PAD in relation to land issues recommended the following:

> In order to promote social harmony and national reconciliation, refugees who left the country more than 10 years ago should not reclaim their original properties because they might have been occupied by other people (PAD 1993).

After the second wave of returnees, Rwanda had no alternative but to opt for land sharing between long-term secondary occupants and returnees, and distribution of state land for new village settlements (Takeuchi and Marara 2009).

Land sharing was introduced with a purpose to ensure that this process is pursued in a fair manner, to maintain peace, harmony, and reconciliation. When land sharing as a process started, this was done by a land committee formed on the very local level (cells) and by collaboration between the head of a cell and the head of a sector (Songo 2014). As an example, if certain properties were repossessed by the displaced people from the 1959 conflict the land committee would mediate and advise that those properties should be shared in such a way that if a house was built before 1959 only half of the property where the house was located remained in possession of the returnee from 1959; the other half went to the returnee from 1994. Land sharing was an informal process; written guidelines for implementation did not exist, and the process was done in different ways in different parts of the country (Potel 2014). Returnees who failed to get land through sharing were resettled on redistributed state land and via village settlement, through a program called *Imidugudu*. Similar to land sharing, allocation of state land for housing and village settlement for returnees was supported by commune chiefs and members, and local authorities (Manirakiza 2014).

With potentially so many overlapping and contradictory claims to land, disputes are likely to appear in post-conflict settings. To prevent these disagreements turning into (renewed) conflict, dispute resolution mechanisms are vital. Depending on the local circumstances, in some cases they might be solved via facilitation and mediation and in others via adjudication. In all cases, land dispute resolution mechanisms are viewed as a necessary tool that must contribute to the peace process (Unruh 2001).

In Rwanda, although almost all returnees were settled one way or another, a number of land disputes were filed to the land claims commission bodies. The first instance was at the local administration authorities. If they failed to settle the land dispute, it was transferred to local arbitration committees, the so-called *Abunzi*. Finally, the highest instance for settling land disputes were the courts (Songo 2014).

In the procedures of solving the land disputes, having land records available appeared to be very important. In general, land records in post-conflict contexts were vulnerable and always negatively affected during the conflict. Land records in post-conflict environments can be damaged, stolen, lost, manipulated by powerful parties (groups), partly or fully destroyed, moved to a third country, or even a target for violent attack (like in 1999 in Timor-Leste). Fully paper-based systems are even more vulnerable (Todorovski 2011). In Rwanda, after the conflict ended it is believed that around 10% of all parcels had formal land records; an example from the Gassabo district shows that only for 7.7% of the parcels land records existed. A small number of existing land records were destroyed or manipulated as a consequence of the conflict, and in the period after the conflict. In about 90% of the territory of Rwanda, customary law was practiced; therefore, informal evidence such as oral witnesses, correspondence letters, and two party sales agreements were socially accepted by both population and authorities. Availability of land records or other proofs of ownership, when existing, appeared to be beneficial for the process of settling land disputes and claims (Manirakiza 2014).

Observing land administration in Rwanda, it is evident that since 1952 private ownership was introduced. In practice, the written law was applied to a small number of people, especially in urban areas, in trading centers, to foreign investors, and for religious institutions. Similar to the colonial period, in the postcolonial period the dual tenure system continued in practice. In the post-conflict period, Rwanda suffered from gaps in land laws and related texts in other land-related laws (Rurangwa 2004). The first developments of post-conflict land administration could be seen with the establishment of the Ministry of Lands in 1999, which was assigned a mandate to develop and draft the new land policy in the period 1999–2003 (Songo 2014). Based on the newly adopted land policy in 2004, Rwanda embarked on a major land reform program, with the perspective that the whole country should be brought under one formal land administration system. Adaptation of the land policy led to the new Organic Land Laws being adopted in 2005 and to the establishment of organizations that had mandate for its enforcement (Potel 2014).

From these developments, as described and evaluated by case studies (Manirakiza 2014; Potel 2014; Songo 2014), we can carefully conclude that in Rwanda the post-conflict context fostered the development of HLP and land administration.

Actors involved in the development of post-conflict land administration in Rwanda can be categorized into four categories: government, citizens, advocacy groups, and international organizations. Each of these actors had their particular role in land administration development. On the government side, there were authorities at the central and local levels. Initially, after the conflict, the Ministry of Lands was formed. Other ministries that became involved at the central level were the Ministry of Agriculture (responsible to administer the rural land in Rwanda), Ministry of Infrastructure, and Ministry of Local Government (the latter two ministries were responsible for administering the urban land in Rwanda). Currently, the Rwanda Natural Resources Agency (RNRA) is responsible for land administration in Rwanda with mandates on central and district levels (Songo 2014).

The Land Tenure Regularization (LTR) program started in 2008, aiming to increase agricultural productivity (through the consolidation of land holdings and greater incentives to invest in agriculture) and hence enhance economic development and reduction of social tension (GoR 2008). Until December 2013, the LTR registered all the land in Rwanda (10.3 million parcels) for the first time. It involved a one-off, low-cost, and community-based process of LTR. The case study by Gillingham and Buckle (2014) identified the following as key success factors: political commitment, a detailed LTR approach developed in an earlier phase of the program, and the program's flexibility. The case study identified some points that needed attention in the longer term: further development of the land administration system, as well as financial and judicial sustainability (Gillingham and Buckle 2014).

Post-conflict state building: the background factors that should be taken into account to understand Rwanda's post-conflict state building are the fact that the RPF gained a complete military victory and that the RPF used the genocide for its own legitimacy building. For states emerging from conflict, initially there are two international norms that support capacity and legitimacy of the state: achieving political stability and the rule of law. In the Rwandan case, political stability was realized by military operations and other institutional arrangements, some of whose contributions to the rule of law were questionable (Takeuchi 2011). The policy of land sharing can be seen as a process, allocating land to the returnees, with the aim of strengthening the political basis of the government. The provision of the housing and village settlement program *Imidugudu* can be understood in the same vein (Hilhorst and van Leeuwen 2000; Huggins 2009). Tackling the sensitive land issues as the Rwandan government did appeared to have a positive influence on the overall security situation in Rwanda, as our case studies show (Songo 2014). It is believed that development and implementation of appropriate land policies has positive influence over the population in support of the legitimacy of the state. (OECD 2010).

According to the case study (Songo 2014), developments in the land sector, as well as in the legal framework, as practiced by state officials, citizen participation, and enforcements by the courts, contributed to the realization of the rule of law in Rwanda. Decentralization of authority and various state development programs (like the LTR program) have had a positive effect on realization of the rule of law in Rwanda (Rugege 2013).

Chesterman et al. (2005) argue that building institutions that are legitimate and sustainable is crucial for developing failing states and states emerging from conflicts. The case results showed that all activities and developments that happened in Rwanda were very much based on the country's specific circumstances: developments in the land sector were not an exception. Building institutions in the land sector was evident in the legal framework, and in the creation of authorities with specific mandates for implementation of the policies and programs (Songo 2014).

The genocide had a very negative social effect on Rwanda, and this was tackled with the promotion of peace, social harmony, and national reconciliation (PAD 1993). Another identified socioeconomic form of legitimacy building from the case of Rwanda is that the state efficiently responded to people's expectations. Takeuchi (2011) acknowledges that post-conflict Rwanda has performed fairly well in terms of providing security and other basic services. Land services could be added to Takeuchi's (2011) acknowledgments: state land was redistributed for housing and village resettlement, and the LTR program provided for whole-of-country land registration. In addition to the improvement of the social sector, high economic growth may have contributed to ameliorating people's living standards (Takeuchi 2011).

Land dispute resolution institutions are also considered to contribute to the overall security situation in post-conflict contexts (Unruh 2001). Land claim commissions and their contributions to the security situation in Rwanda have been mentioned previously.

Discussion

This section focuses on identifying the characteristics of post-conflict state building that relate to HLP and land administration. First, the background factors of land, conflict, and post-conflict context are covered. Following this, elaboration of the characteristics of post-conflict state building is undertaken with focus on the land. This is done in using structure of the characteristics of war-torn societies (Ball 2001). The aim is to illustrate and increase the understanding of HLP and land administration and how they developed in post-conflict state building. The findings from the Rwandan case are used. A table of characteristics of post-conflict state building, focusing on HLP and land administration, is finally presented.

Background factors of land, conflict, and post-conflict context: causes of violent conflicts are many and varied. In the case of Rwanda, land issues were considered as a major cause that was used to increase ethnic divisions, leading to violent conflict (Magnarella 2005). This is an important fact to be considered, because land therefore requires attention in post-conflict state-building activities, as was the case in Rwanda. The most serious concerns from a violent conflict are number of people killed, destruction of properties and infrastructure, and displacement. The conflict in Rwanda ended with a genocide in which one million people died (Prunier 1997) but without substantial destruction of houses and infrastructure. The cases of Kosovo and Timor-Leste are examples where PADs contained specific land management and land administration activities integrated in the post-conflict UN operations (AD 1999; UN 1999). In a majority of other cases, unfortunately in the case of Rwanda (Bruce 2007) as well, PADs had very limited attention to land. This could be a reason for later developments that occurred in the reconstruction phase of the post-conflict period. All violent conflicts result in displacement of population (Rwanda had more than two and a half million refugees), which is an alarming issue for the host countries, the UN, and the international community. Displacement and destruction of properties have large impacts on land and its administration (Hollingsworth 2014): together, they are a critical issue in recent conflict and post-conflict environments.

In the early stages of the post-conflict period in Rwanda, political stability was realized with military intervention (Takeuchi 2011). Taking a longer term view, developments of HLP and land administration contribute to political stability integrated in the broader National Development Strategy. Recognizing that the rule of law is critical to states emerging from a conflict (Rugege 2013), it was found that the developments of legal framework, establishment of organizations, and participation of the citizens in the land sector (Songo 2014) contributed to the realization of the rule of law in the case of Rwanda.

Institutional weaknesses: in post-conflict contexts, states face institutional weaknesses like limited legitimacy of the state, a nonfunctioning political system and government structure, and an inadequate legal framework. In Rwanda, it was found that political stability initially was based on the land sharing policy, allocation of state land, and housing and village settlements (Hilhorst and van Leeuwen 2000; Huggins 2009). Development and implementation of various policies, including land policies, is considered to have a large influence over the population, in the sense of strengthening the legitimacy of the state (OECD 2010). Establishment of organizations such as land claim commissions, different ministries responsible for land, and the RNRA in Rwanda as governmental entities contributed to the post-conflict government structure. The example of Rwanda shows that first a land policy was developed, which later led to the development and adoption of land laws and the appointment of a specific organization with a mandate for enforcement of the laws and land policy (Songo 2014). From this, we can derive that strengthening the institutional weaknesses in the land sector is clearly important for Rwanda.

Social and economic problems: post-conflict contexts witness social and economic problems such as death and injury, displacement, destroyed properties and infrastructure, hatred between ethnic groups, loss of proof of identity, and ownership. The genocide in Rwanda created a very negative social situation, which was addressed by the promotion of peace, harmony, and national reconciliation. Addressing land issues in the PAD (1993) in a context of social harmony and national reconciliation was a first step of involvement of HLP and land administration in the improvement of the social and economic situation. This commitment, from the PAD, was further identified when local authorities supported the displaced people and assisted in the land sharing and village settlement processes. Where it was required, sector authorities mediated and advised on the land sharing and distribution of state land between returnees from different periods. This mediation is seen as a successful land dispute tool, and if they failed then the official three-level land claim committees became involved (Potel 2014). Development of the land policy and land law resulted in the improvement of the land registration and land administration system in Rwanda. Development of land administration in Rwanda was supported by the LTR program, which contributed to sustainable development and supported the real property market. Initial success of the LTR program suggests that it influenced an increase in agricultural productivity and economic development and reduced social tensions. Development of land administration contributed to the service provision to all land-related sectors and significantly increased the security of land rights (Gillingham and Buckle 2014). This is also seen as a support of the establishment of the land market within the overall economic development of Rwanda. Acknowledging developments in HLP and land administration, which contributed to the development of land-related sectors, leads to their identification as elements of social and economic importance in post-conflict Rwanda.

Security problems: the first challenge that post-conflict states face is the process of keeping the peace (Call and Cousens 2008). The security situation in post-conflict Rwanda could be described as tense and fragile, with a large military and UN presence (Prunier 1997; Takeuchi 2011). Land dispute resolution mechanisms, widely viewed as legitimate and pursued by the state, in general can contribute to a peace process (Unruh 2001). In Rwanda, mediation and advice by sector authorities on the land sharing and distribution of state land, supported by settling land disputes, significantly reduced conflict tensions over land. The three-level land claim committee bodies were available as official authorities in Rwanda (Manirakiza 2014). Addressing land in

TABLE 5.1

Characteristics of Post-Conflict State Building Focusing on HLP and Land Administration in Rwanda

Institutional Weaknesses	Economic and Social Problems	Security Problems
Legitimacy of the state (strengthen by land policies) Political stability (supported with land policies) Government structure (authorities dealing with land claims, land administration) Legal framework (land policy, land law, land registration)	Land issues in context of peace, social harmony, and national reconciliations Displacement–land relation (land sharing, state land, housing and village settlements) Community participation Citizen participation Implementation programs (with aim to develop agriculture, reduce social tensions, and improve economic situations)	Land claim commissions (mediation or adjudication method) Including land in PAD (to some degree supports security situation as well)

the PAD (1993), facilitating land redistribution, land sharing and the village settlement program by the community members and local authorities, and resolving land disputes and claims (Potel 2014) could therefore be identified as elements of the security situation in post-conflict Rwanda.

As a synthesis of the discussion, Table 5.1 illustrates the identified elements of HLP and land administration in Rwanda and their fit within overall post-conflict state building.

Conclusions

This chapter aimed at recognizing the characteristics of post-conflict state building, with a focus on HLP and land administration. This was done initially by identifying the background land-conflict factors and then using the characteristics of war-torn societies (Ball 2001) as a framework for understanding the observations made on HLP and land administration within post-conflict state building. Based on the contributions from literature and the presented findings for Rwanda, the following generalized conclusions are made.

As a background factor, land can be one of the many causes for a conflict, but in some cases land is considered the major cause for violent conflict. Therefore, land must always be considered as part of post-conflict state building activities. Addressing land in general, and specifically land management and land administration, in a PAD, and within UN operations, is beneficial, and it can result in improvements to land administration in earlier phases of the post-conflict period. However, HLP and land administration tend to contribute to political stability in the longer term and should be integrated in the broader National Development Strategy. Development of the legal framework, practiced by state officials, with citizen participation, and with enforcement of the law via courts, promotes the rule of law in post-conflict contexts. Similar developments in the land sector also contribute to this promotion of the rule of law.

One of the characteristics of post-conflict state building is institutional weaknesses, which could initially be addressed by applying specific land policies. Land policies can have significant influence over the population, and they contribute to creating legitimacy for the post-conflict state. Usually, a ministry of lands or similar governmental authority has a mandate to develop the land policy. Land policies lead to the development of land laws and indicate organizations for enforcement of the policies and the laws. Examples are land administration organizations and land claim commissions within the governmental structure. Implementation of legal frameworks is most successful via appropriate implementation programs.

Another characteristic of post-conflict state-building activities is ongoing social and economic problems, where the most negative social effects are death and injury. These problems can be tackled with promotion of peace, social harmony, and reconciliation. Land-related issues, specifically in relation to a displaced population, can be addressed in the same manner. Mediation and advice in allocation of land to returnees, or land policies supported by local authorities, and official land claim commissions for settling land disputes can support the social situation. Improvement of the service provision, within the land administration sector, via appropriate implementation programs increases the security in land rights. A supportive implementation program in land administration can support the development of a land market and the overall economic situation of a post-conflict state.

By definition, the security situation is always likely to be problematic in a post-conflict context. Land dispute resolution institutions are accepted as contributors to overall security in a post-conflict situation. Including land dispute resolution institutions in the PAD and in the UN operational guides can speed up the development of such institutions. Availability of mediation methods or official land dispute resolution institutions is recognized as an element that supports the security situation after the conflict.

Based on the findings of this study, it is concluded that the potential of HLP and land administration has been identified, and indeed both play a particular role in post-conflict state building. Table 5.1 shows the characteristics of post-conflict state building and how HLP and land administration are applied for a specific context. The table can be applied in other cases and illustrates the potential of HLP and land administration in other post-conflict state-building cases.

References

AD. 1999. Agreement between the Republic of Indonesia and the Portuguese Republic on the Question of East Timor. New York, USA: United Nations.

Ball, N. 2001. "The challenge of rebuilding war-torn societies." In *Turbulent Peace*, edited by A. Croker, Hampson, F. O., and Aall, P. Washington, DC: U.S. Institute of Peace.

Brahimi, L. 2007. State building in crisis and post-conflict countries. 7th Global Forum on Reinventing Government Building Trust in Government. Vienna, Austria: United Nations.

Bruce, J. 2007. "Drawing a line under the crisis: Reconciling returnee land access and security in post conflict Rwanda." In *HPG Working Papers*. London, United Kingdom: Overseas Development Institute.

Call, C. T., and E. M Cousens. 2008. "Ending wars and building peace: International responses to war-torn societies." *International Studies Perspectives* 9(1):1–21.

Chesterman, S., M. Ignatieff, and R. Thakur. 2005. *Making States Work: States Failure and the Crisis Government*. Tokyo, Japan; New York; Paris, France: United Nations Press.

de Waal, A. 2009. "Why humanitarian organizations need to tackle land issues." In *Uncharted Territory: Land, Conflict and Humanitarian Actions*, edited by S. Pantuliano. London, United Kingdom: Overseas Development Institute.

EU-UN. 2012. Land and Conflict. In *EU-UN Partnership: Toolkit and guidance for preventing and managing land and natural resources conflict*. New York, USA: UN Interagency Framework Team for Preventive Action.

FAO. 2005. "Access to rural land and land administration after violent conflicts." *In FAO Land Tenure Studies*. Rome, Italy: FAO.

Fukuyama, F. 2004. *Statebuilding: Governance and World Order in the 21st Century*. Ithaca, NY: Cornell University Press.

Guterras, A. 2009. "Foreword " In *Uncharted Territory: Land, Conflict and Humanitarian Actions*, edited by Sara Pantuliano. London, United Kingdom: Overseas Development Institute.

Gillingham, P., and F. Buckle. 2014. Rwanda Land Tenure Regularisation Case Study. Hertfordshire, United Kingdom: HTSPE Limited for Evidence on Demand and UK Department for International Development (DFID).

GoR. 2008. Strategic Road Map for Land Tenure Reform in Rwanda, Government of Rwanda. Kigali, Rwanda: Government of Rwanda.

Heinz-Jürgen, A, A. Milososki, and O. Schwarz. 2006. Conflict—A Literature Review. Duisburg, Germany: Department of Social Sciences, Institute for Political Sciences, Jean Monnet Group, Universitat Duisburg Essen.

Hilhorst, D., and M. van Leeuwen. 2000. "Emergency and development: The case of Imidugudu, villagization in Rwanda." *Journal of Refugee Studies*, 13(3):264–280.

Hollingsworth, C. 2014. *A Framework for Assessing Security of Tenure in Post-Conflict Contexts*. MSc theses. Enschede, The Netherlands: Faculty ITC, University of Twente.

Huggins, C. 2009. "Land in return, reintegration and recovery processes: Some lessons from the Great Lakes region in Africa." In *Uncharted Territory: Land, Conflict and Humanitarian Actions*, edited by S. Pantuliano. Overseas Development Institute. London, United Kingdom.

Kato, H. 2014. "Foreword of *Confronting Land and Property Problems for Peace*." In *Confronting Land and Property Problems for Peace*, edited by S. Takeuchi. London, United Kingdom: Routledge.

Leckie, S. 2000. *Resolving Kosovo's Housing Crisis: Challenges for the UN Housing and Property Directorate. Forced Migration Review* (7):12–16. [accessed date January 31, 2014]. Available from http://www.fmreview.org/en/FMRpdfs/FMR07/fmr7full.pdf.

Lewis, D. 2004. "Challenges to Sustainable Peace: Land Disputes Following Conflict." FIG Symposium on Land Administration in Post Conflict Areas. Geneva, Switzerland.

Magnarella, P. 2005. "The background and causes of the genocide in Rwanda." *Journal of International Criminal Justice*, 3(4):801–822.

Manirakiza, J. G. 2014. *The Role of Land Records in Support of Post-Conflict Land Administration: A Case Study of Rwanda in Gasabo District*. MSc theses. Enschede, The Netherlands: Faculty ITC, University of Twente.

Murekezi, A. 2012. *Rebuilding after Conflict and Strengthening Fragile States: A View from Rwanda*. Vol. 22. Harare, Zimbabwe: African Capacity Building Foundation (ACBF).

Nyamwasa, K., P. Karegeya, T. Rudasingwa, and G. Gahina. 2010. Rwanda Briefing. Kigali, Rwanda: Africa Faith and Justice Network.

OCHA. 2005. United Nations Humanitarian Response Review. New York; Geneva, Switzerland: Office for the Coordination of Humanitarian Affairs.

OECD. 2008. Concepts and Dilemmas of State Building in Fragile Situations, from Fragility to Resilience. Paris, France: Organisation for Economic Co-operation and Development.

OECD. 2010. The State's Legitimacy in Fragile Situations: Unpacking Complexities. Paris, France: Organisation for Economic Co-operation and Development.

PAD. 1993. Peace Agreement between the Government of the Republic of Rwanda and the Rwandese Patriotic Front. Arusha, Tanzania: United Nations.

Pantuliano, S. 2009. "Uncharted territory: Land, conflict and humanitarian actions." In *Uncharted Territory: Land, Conflict and Humanitarian Actions*, edited by S. Pantuliano. London, United Kingdom: Overseas Development Institute.

Potel, J. 2014. *Displacement and Land Administration in Post-Conflict Areas—Case of Rwanda*. MSc theses. Enschede, The Netherlands: Faculty ITC, University of Twente.

Prunier, G. 1997. *The Rwanda Crisis: History of a Genocide*. New York, USA: Columbia University Press.

Putzel, J. 2009. "Land governance in support of the millennium development goals, a new agenda for land professionals." In *FIG Publication 45*. Washington DC, USA: FIG/World Bank.

Rugege, J. S. 2013. "Judicial reform, public confidence and the rule of law in Rwanda." In *Key Note Speech, Qatar Law Forum*. London, United Kingdom: Qatar Law Forum.

Rurangwa, E. 2004. Land administration in post conflict situation-Rwanda case. In *Proceeding of FIG Commission 7 Symposium on Land Administration in Post Conflict Areas*, edited by P. van der Molen and Lemmen, C. Geneva, Switzerland: FIG.

Songo, M. N. 2014. *Roles of Actors in Early Recovery Post-Conflict Land Administration in Rwanda: Rationale for Guidelines Improvement*. MSc theses. Enschede, The Netherlands: Faculty ITC, University of Twente.

Takeuchi, S. 2011. "Gacaca and DDR: The diputable record of state-building in Rwanda." In *State-Building in Conflict-Prone Countries*. Tokyo, Japan: JICA Research Institute.

Takeuchi, S. 2014. *Confronting Land and Property Problems for Peace*, edited by S. Takeuchi. London, United Kingdom: Routledge.

Takeuchi, S., and J. Marara. 2009. *Conflict and Land Tenure in Rwanda*. Tokyo, Japan: JICA Research Institute.

Todorovski, D. 2011. "Characteristics of Post-Conflict Land Administration with Focus on the Status of Land Records in Such Environment." FIG Working Week: Bridging the Gap between Cultures. Marrakesh, Morocco.

Todorovski, D., J. Zevenbergen, and P. van der Molen. 2012. "Can Land Administration in Post-Conflict Environment Facilitate the Post-Conflict State Building?—a Research Problem." FIG Working Week: Knowing to Manage the Territory, Protect the Environment, Evaluate the Cultural Heritages. Rome, Italy.

UN. 1999. "Report of the secretary-general on the United Nations interim administration mission in Kosovo." New York, USA: United Nations.

UN/ECE. 1996. *Land Administration Guidelines*. New York; Geneva, Switzerland: United Nations Economic Commission for Europe.

UNHCR. 2014. *UNHCR Global Trends 2013—War's Human Costs*. UNHCR 2014 [accessed date August 4, 2014]. Available from http://www.unhcr.org/5399a14f9.html.

UN-HABITAT. 2007. *A Post-Conflict Land Administration and Peacebuilding Handbook*. Nairobi, Kenya: UN-HABITAT.

UN-HABITAT. 2009. *Land and Conflict; Handbook for Humanitarians*. Nairobi, Kenya: UN-HABITAT.

Unruh, J. 2001. "Postwar land dispute resolution: Land tenure and peace process in Mozambique." *International Journal of World Peace*, 18(3):23.

Unruh, J., and R. Williams. 2013. "Lessons learned in land tenure and natural resource management in post-conflict societies." In *Land and Post-Conflict Peacebuilding*, edited by J. Unruh and Williams, R.C. London, United Kingdom: Earthscan.

van der Molen, P. 2002. "The dynamic aspect of land administration: An often-forgotten component in system design." *Computers, Environment and Urban Systems*, 26(5):361–381.

Wehrmann, B. 2005. *Urban and Peri-Urban Land Conflicts in Developing Countries*. Phd theses. Urban and Regional Geography. Marburg, Germany: Philipps-Universität Marburg.

Yin, R. K. 2003. *Case-Study Research—Design and Methods. Applied Social Research Methods Series*. Third edition. Vol. 5. Thousand Oaks, CA: Sage Publications.

Zevenbergen, J., and T Burns. 2010. "Land Administration in Post-Conflict Areas; a Key Land and Conflict Issue." 24th Paper read at FIG Congress: Facing the Challenges—Building the Capacities. Sydney, Australia.

6

Land Administration Crowds, Clouds, and the State

Peter Laarakker, Jaap Zevenbergen, and Yola Georgiadou

CONTENTS

Introduction

In his book *Seeing Like a State*, Scott (1998) describes the introduction of cadastre, civil registry, and other systems in western Europe at the beginning of the nineteenth century. The purpose of that introduction was the creation of legibility of society so that the emerging state could administer its territory and population. The legibility created by states was not identical to the existing social reality; the state had to limit itself in the collection of information, and any description was by definition an abstraction. Scott describes how the state information replaced social reality because the state has the power to force it on society by denying services to citizens if those are not requested for in terms of this state-created reality.

Contemporary research on land registration systems in the developing world is placed against a much more complex societal background. Next to the state and the citizen in the dichotomous relationship that Scott describes, many more actors are involved. Land rights' claimants organize themselves in groups often with the support of national or international nongovernmental organizations (NGOs), whereas donors and multilaterals and their knowledge centers influence the processes. Nowadays, states are more restricted in their actions through their participation in international agreements with other states or lending organizations such as the World Bank. There is also attention on obscure actors that are influencing the processes through activities in general negatively labeled as corruption, discrimination, or land grabbing.

In this complex international context, state-run land registration systems are supplemented or even challenged by "crowds" and "clouds." The concept of crowds derives its meaning from the process of crowdsourcing: "Using many contributors without regard to their background and skill level" (Haklay et al. 2014). Crowdsourcing of geographic information in general is known as Volunteered Geographic Information (Goodchild 2009; Goodchild and Glennon 2010), and it is intensively researched. Participants of such activities are often referred to as human or citizen sensors (Georgiadou et al. 2014). Crowdsourcing of (geographic) information that is relevant in the context of land registration comes under many different concepts such as community-driven adjudication, grassroots mapping, participatory (resource) mapping, and social data gathering, among others (Laarakker and De Vries 2011; De Vries et al. 2014). These concepts represent practices where landowners or claimants themselves initiate the recording of their rights or claims. In such cases, crowds are not anonymous since the individual actors often have a very personal interest: to articulate their rights or claims.

Initiatives for these participatory forms of data collection can be taken either by the state itself—driven by good governance or efficiency principles—or without state involvement. In the latter cases, information on land rights or claims is produced to make such claims locally consistent and explicit so that they can be used for the interrelated land affairs of the community that performs the data collection or in the negotiations of that community with the state. Very often, national or international NGOs are involved. Their perspectives on land rights obviously influence the outcome of the process as well.

Clouds are closely connected to crowds. The concept of clouds in land administration stands for practices that make information on land rights and claims and other land registration-related information public on the Internet. Clouds and cloud computing stand for a new paradigm for the provision of computing infrastructure and information and communications technology (ICT) services (Vaquero et al. 2009). Vaquero et al. analyze a large number of different definitions of cloud and synthesize these to a large pool of easily usable and accessible virtualized resources that are scalable and often have a pay-per-use utility model. In that way, infrastructure, platforms, and software can all be provided as a service. The organizations that operate cloud services in the domain of land registration aim to secure land and resource rights by promoting transparency in decision making over land and investment (Land Matrix 2013) and by posting recorded evidence on land rights and claims on a global, secure, open, and free platform (Community Land Rights 2014; McLaren 2014).

The advancement and mutual enforcement of crowds and clouds are facilitated by concepts such as the "continuum of land rights" (GLTN 2008) and the concept of legitimate tenure in the Voluntary Guidelines (UN-FAO 2012) that acknowledge both the existence of land rights and the forms of tenure outside state-run land registration systems. Developments in geo-ICT provide not only the technical tools needed to allow the crowd to document their land rights and claims but also the models to manage these data and publish them on the web (Lemmen et al. 2007; McLaren 2011).

Information collected by crowdsourced methods does not necessarily lead to changes in state-run systems. The information can be stored in alternative cooperative systems that are acknowledged by the state (Bennett et al. 2012). However, when the information is not acknowledged by the state, the data set built by the crowd is contradicting the state-run system. De Vries et al. (2014) qualify the so-produced neo-cadastres as an "artefact of dissatisfaction."

Concerns have been expressed about the legitimacy of the information produced by crowdsourced methods (McLaren 2011). While state-run systems may contain illegitimate information because of corruption, discrimination, or land grabbing, to name a few causes, crowdsourced systems are also vulnerable to similar defects. Despite the belief that more participative methods of land registration better reflect social reality than less participative ones, it is important to realize that participative registration is also an abstraction of reality. Therefore, knowing how "participative abstraction" takes place is important as well.

We have to understand better what is happening when crowds and clouds are entering the scene. How do they operate, and what is the result of their endeavors? Are crowds and clouds doing a "better" job than the state, and from which economic, legal, or ethical perspective should we define what is better? Less state is more citizen participation but also less guidance and standardization. Can crowds and clouds counter practices like land grabbing, discrimination, and corruption; can we objectively define such practices; and are these new approaches themselves free from such influences?

The approach of this chapter is to have a more neutral position and to develop a conceptual framework with which the authority of both state and non-state actors can be assessed. In his closing speech at the World Bank land and poverty conference (2014), Mahmoud Mohieldin, the special envoy for the president of the World Bank, urged not to look at the world through rose-tinted glasses: "The world is full of bad ideas and bad habits and many people are benefitting from that." Models for land registration systems are normative and imbued with values, often not explicitly stated. They assume that the bad world can be kept outside. This research has the objective to develop a model for land registration systems and their innovative challengers that accommodates both the legitimate and the non-legitimate actions of all actors in building up such systems.

In the section "Theoretical Perspective," the existing theoretical framework for land registration is discussed. The existing framework is not able to fully accommodate the introduction of crowds and clouds. Grounded theory, the methodology chosen for this research, is motivated in the section "Methodology." In the section "Results," the results from a literature study using the methodology of grounded theory are given. In the section "Discussion," the discussion of the results leads to an enhanced conceptual framework.

Theoretical Perspective

Crowds and clouds are involved in many aspects of land administration and land management. This chapter focuses on their contribution to land registration. McLaughlin and Nichols (1989, p. 81) define land registration as

> the process of recording legally recognised interests (ownership and/or use) in land.

Nichols and McLaughlin (1990, p. 107) define land registration slightly differently:

> the process of officially recording information about land tenure. Land tenure encompasses the rights, responsibilities and restraints that govern the allocation, use and enjoyment of land resources.

The land records that are the result of the process of land registration are illustrated by Henssen (1995) with the classical trinity man–right–parcel. The first definition takes "legally recognised interests" as the starting point. The second refers to "officially recording information about land tenure." Concepts such as "legally" and "officially" refer to a certain state authority. The concept of land registration in the quoted definitions and in Henssen's model presupposes a state that has the authority to run the land registration system. Figure 6.1 makes this explicit: the rights relationship between a person or group of persons and a defined unit of land is defined by the state (Laarakker et al. 2014). The authority of the state concerns the natural or legal persons who can have a right relationship to land, units of land that can exist, types of right relationships that can be registered, and specific rights

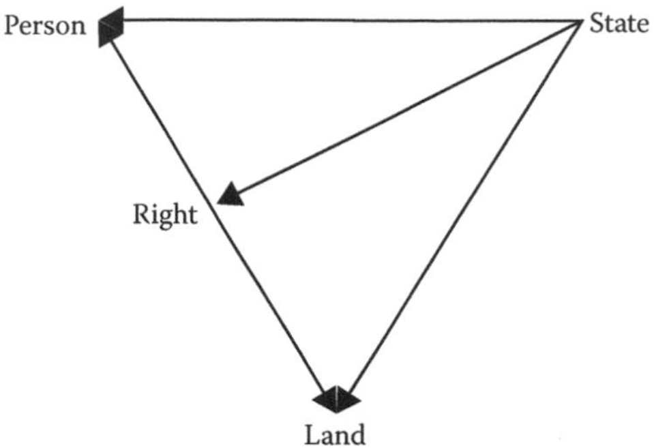

FIGURE 6.1
The state defines the person–land relationship.

relationship between a defined person or group of persons and a defined unit of land. Right should be seen as the sum of all rights, restrictions, and responsibilities (Bennett 2007) a person has.

A sovereign state has a number of attributes: a permanent population, a defined territory, government, and the capacity to enter into relations with other states (Seventh International Conference of American States 1933). The introduction of the state in the model reflects the analysis by Scott (1998) that the state administers a land registration system to make society legible. In that way, the state cannot only control society (taxation, enlist people for the army, etc.) but also develop society (better health care, etc.). Literature on land registration does not reflect much on the control aspect of the state in land registration. The aspect of development, however, is well covered in literature by De Soto (2000) and other authors who discuss his ideas (Gilbert 2002; Von Benda-Beckmann 2003; Barros 2010). For example, the land management paradigm (Williamson et al. 2010) advances the belief that the introduction of land administration systems (of which land registration is a crucial part) will lead to sustainable development.

Each state's own world view will lead to a specific definition of sustainable development (Van Egmond and De Vries 2011), which will affect the design of the land registration system including the role of the state within that system. The involvement of states in land registration systems differs considerably around the world. There is very low involvement in certain states of the United States (Larsson 1991, pp. 52–56) or in countries with a weak government where for practical reasons the state is not involved. There is also an ideological difference between deeds registration systems, where the bilateral agreement between two parties is decisive in the creation of land rights, and title registration systems, where the registration of a title by the government is decisive. Zevenbergen (2002, p. 60) finds that convictions on the title–deed discussion have even the characteristics of religious beliefs. The political ideology in a country will also determine the view on crowds and clouds. For these reasons, the previously absent or taken-for-granted state is now an explicit variable in the model in Figure 6.1.

Besides the state that runs a land registration system, non-state actors also develop, administer, and maintain land recordings on person–land relationships. We define land recording as any conceptualization of person–land relationships that can be enforced in a certain social context. This definition includes state-run land registration systems, locally used paper documentation that is referred to by Zevenbergen et al. (2013, p. 598), and information sets built up by indigenous communities supported by NGOs (Di Gessa 2008; Flintan 2012; Ramirez-Gomez et al. 2013). Literature also describes situations where no written documentation exists and where land is governed and conflicts are resolved by oral rulings that are not written down. Van Leeuwen (2014) refers to "communal memory." Graeber (2007) refers to "a broader sense of justice" and "traditional principles" as a source for local rulings that are not in conformity with the state-run registration system. Such communal memory, sense of justice, and traditional principles fall under the concept of social institutions as defined by Simbizi et al. (2014). All must somehow conceptualize person–land relationships since they aim to lead to decisions in this field. We include these mental recordings in the definition of land recording.

For the same geographical area, conflicting recordings can exist (Simbizi et al. 2014). Non-state actors can claim legitimacy for their recordings while declaring the state-run registration illegitimate.

Tools are provided for such situations. The purpose of the Community Land Rights initiative (Community Land Rights 2014) is to produce an online map with land claims. Barry and Asiedu (2014) promote the Talking Titler approach if the official register is not modeling the de facto situation on the ground adequately. The Social Tenure Domain Model provides a data model for land recordation that accommodates extra-legal rights (Griffith-Charles 2011) and overlapping claims (Lemmen et al. 2007; Lemmen 2012, pp. 131–149). So, crowds and clouds are facilitated to produce land information that sometimes lacks the "official" or "legally recognized" status but still gets recognition in social contexts other than the state and so can be more than only an artifact of dissatisfaction. All this runs counter to the definitions in the beginning of this section by McLaughlin and Nichols (1989), Nichols and McLaughlin (1990), and Henssen (1995). It is therefore important to expand the traditional theoretical perspective on land recordation to understand the contribution of crowds and clouds to it.

Methodology

The objective of this chapter is to have a look at land recordation in its widest sense and develop an overall theoretical framework that accommodates all land recordations as defined in the former section and the contribution of crowds and clouds to these processes. To develop such a framework, a wide array of existing theories can be used. Besides the classical sciences such as technology and law, political and social sciences, anthropology of law and state, and public administration can also contribute to the understanding of these processes.

The phenomena of crowds and clouds in the field of land registration are rather new and not yet systematically studied. The choice of one particular theoretical lens to study these phenomena creates the risk of missing out relevant aspects. For this reason, a grounded theory methodology as developed by Strauss and Corbin (1990, 1998) was chosen. Grounded theory aims at inductive theory development within a positivist paradigm (Ozanne 1992). The essence of the methodology is that the theory is being developed during the research in a hermeneutic process (Verschuren and Doorewaard 1999). The data are taken as a starting point for the research, and the research eventually leads to a theory that is fit for its supposed use.

Strauss and Corbin acknowledge that the researcher is not a blank slate. He or she is influenced by his or her background and prior theoretical and empirical knowledge before studying the actual subject, in this case crowds and clouds. This prior knowledge will have an effect during the development of the grounded theory and prohibits an unbiased view of the phenomena to be researched. Existing theory was therefore given a place in the methodology. The developed methodology combines an open-minded approach of the subject in the first part of the research with an assessment of usability of existing theory in the second part. In such a way, the grounded theory methodology can also be used for the enhancement and adaptation of existing theories (Strauss and Corbin 1990, 1998).

Karsten and Timmers (2008) list a number of advantages and disadvantages for the use of existing literature. As a disadvantage, they name the risk of missing aspects and suppressing insights, as well as biased selection and interpretation of cases. However, the advantages are that literature gives focus, enlarges the insight in the subject and the theoretical sensitivity of the researcher, and gives a better connection to existing knowledge.

Wolfswinkel et al. (2013) have also used a grounded theory approach for the execution of a literature review to reach a thoroughly and theoretically relevant analysis of the topic of the review. Since this research consists of a review of existing literature in a multi-theory environment, the methodology seems fit to be used.

Following the methodology developed by Wolfswinkel et al. (2013), literature on land rights registration was selected based on key words. Initial key words were participatory mapping, crowdsourcing, and grassroots mapping. Several empirical cases are described in the gray literature such as publications from the United Nations Habitat, Global Land Tool Network (GLTN), and NGOs. Some journal articles were also selected. The case descriptions were coded and categories were developed. The coding of cases was done in an iterative way and in parallel to an analysis of existing scientific literature.

Results

In this section, the main categories emerging from the studied cases are discussed. In each subsection "Reflections," the results of the coding are compared with findings from scientific literature.

Taking the Initiative

In the case descriptions that make clear how the process of recordation started, a point of initiative can be distinguished (Flintan 2012). Sometimes, it is the state itself that with the objective of greater participation or more efficiency develops methods with higher involvement of local people. In Rwanda (Lemmen and Haarsma 2012), the Agency for Natural Resources trained local para-surveyors in the framework of a national land registration project. In other described cases, local communities take the initiative. Patel and Baptist (2012) describe many community-led mapping and documentation projects. In the case described by Ramirez-Gomez et al. (2013), the initiative is taken by an NGO. In most of these projects, NGOs are supporting the local communities. The studied land registration clouds are all led by NGOs (Land Matrix 2013; Community Land Rights 2014; McLaren 2014). Some cases also show a mixture of state and non-state actors that take the initiative (Flintan 2012).

Initiatives can be grouped into those that are led by the state and those that are led by non-state actors. Non-state-led initiatives vary in the degree to which they take existing state governance as a starting point. Information can be collected with the objective to submit the data as soon as possible to the government (Archer et al. 2012). In other cases, information is collected in anticipation of a longer battle with the government over the claimed land rights.

Reflections

The original key word "participatory mapping" appears to be a concept that is always used in relation to state power. In the ladder of participation developed by Arnstein (1969), *citizen control* is the highest level of participation. This level of participation still requires a state for final approval and accountability. Non-state-led initiatives differ in whether they start from a participation paradigm in which the state is taken as a reference or a more activist paradigm in which the state is seen more as an opponent. The latter is dealt with in the section "State-Challencing Attitude." Initiative is a category, but it should always be looked at in the context of the long-term objectives of the taker of the initiative.

Deconstructed State

In the studied cases, there is often reference to legal recognition by the state of rights in situations in which it is still problematic to get protection for these rights. Legal recognition by the state takes many forms: recognition of a particular man–land relationship by accepting registration in a register held by the state, recognition in a civil code or other statutory law that a certain right can exist or can be registered, as well as recognition on a more general legal level in the constitution. The state also establishes recognition of tenure relations by signing multilateral or bilateral international treaties or an obligatory agreement with a donor organization. Royston and Du Plessis (2014) also distinguish administrative recognition, a situation in which the tenure relation as such is not legally recognized but access to public utilities is awarded by administrative officials. Singh (2014) describes similar practices in Bhopal, India. Bakker (2006) describes the ambivalent position of adat—customary rights in Indonesia. The adat is recognized by the Basic Agrarian Law of Indonesia, but no operational law is in place to register these rights. De Vries and Zevenbergen (2011) describe conflicting regulations defining discretional space of land registration officials. UN-FAO (2012, p. 29) refers to "competing rights" in Principle 17.2. Holston (2008) describes the very complicated relationship between the state-run land registration system and the de facto tenure situation in Sao Paulo, Brazil. In such cases, "recognition" becomes a fuzzy concept in need of reference to a specific law. In many cases, these sources of recognition of rights appear not to be consistent; the deconstructed state emerges as a category.

Reflections

The concept of a deconstructed state is in accordance with Li (2005), who contests the state as a monolithic entity with one intention and as a container of power (Richter and Georgiadou 2014, p. 3). A conceptual framework for land registration should acknowledge this multipolarity of the state. Harvey (2006) uses the concept of elasticity between land tenure and cadastral registration to describe the situation in post-communism Poland, where the tenure situation in reality is to a very limited extent reflected in the cadastral records but where no cost-effective procedures are available to repair this and so the situation is accepted by certain entities of the state. Zevenbergen et al. (2013, p. 599) are aware of potential inconsistencies between state entities when underlining that roles and responsibilities have to be made very clear when involving local communities in the process of land recordation. Apparently, this is not always done.

Figure 6.2 displays the concept of the deconstructed state as a further enhancement of the model presented in Figure 6.1. Different actors inside the state structure can qualify the same person–land relationship differently.

With this extended conceptual framework, we apply a more detailed level of analysis than the macro level used by Scott (1998) to look at land registration systems. Later studies criticized his dichotomy state–citizen dichotomy. In a study on land registration in Indian cities, Richter and Georgiadou (2014) found proof for the idea of Li (2005) that state–citizen is not a dichotomy but that state and citizen might overflow. In the model in Figure 6.2, this means that a multitude of relations between persons and actors in the state can exist, with specific levels of overflow between them, leading to different assessments of the rights relation of a specified unit of land. In this way, the model also accommodates corrupt and discriminatory practices.

The concept of a deconstructed state can be found implicitly in the internationally accepted continuum of land rights (GLTN 2008). The continuum is meant to show the range of possible forms of tenure. The continuum arranges land rights from formal to informal, reflecting a decreasing level of state recognition, registration, and protection. The more the entities of a deconstructed state recognize a certain right and the related registration, the more legible this right and the more formal this right. Royston and Du Plessis (2014, p. 6) quote Fourie, stating,

> a Continuum of Land Rights [...] can be said to exist if a land information management system includes information that covers the whole spectrum of formal, informal and customary rights.

If the land information management system is state run, each right in the system will be based on the recognition by one or more specific entities of the deconstructed state. It has to be clear which information has to pass by which state-appointed gatekeeper (compare the work by GLTN et al. [2012, p. 36]). If the different parts of the deconstructed state produce conflicting

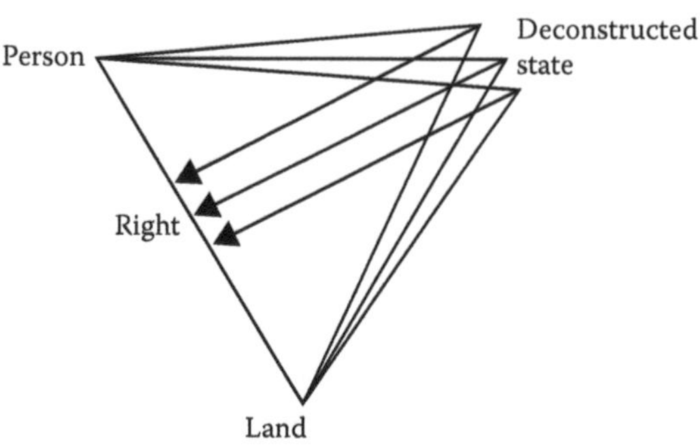

FIGURE 6.2
The deconstructed state.

rights and both these conflicting rights are accepted by the land information management system, a consistent procedure has to be designed to solve these conflicts. Van Leeuwen (2014) even describes a situation of conflicting conflict solution procedures.

For land information management systems that are not state run, see the section "State-Challenging Attitude."

Recordation Vocabulary

Initially, literature was selected that describes case studies in which a legal vocabulary (GLTN 2008; McCall 2011) was used in the application of crowdsourcing methods. Several case studies were found that describe processes of so-called participatory resource mapping (Flintan 2012; Ramirez-Gomez et al. 2013) where people were mapping the areas on which they are dependent for their livelihood, using a different vocabulary. The resources were mapped to collect the information to settle land issues between communities or between communities and the state. In such situations, mapping cannot be seen as an objective registration of de facto dependencies but must also be seen as a preliminary stage for claiming land rights (Ramirez-Gomez et al. 2013).

Another set of relevant cases deals with participatory planning (McCall and Dunn 2012). Most of the time, spatial planning will lead to a restriction on the use of land, so participation in planning processes is affecting rights of individuals. In these descriptions, a different set of concepts is used to make inventories. Patel and Baptist (2012) use the term enumeration for collecting census information in informal settlements. They distinguish three forms of information gathering for informal settlements: profiling, enumeration, and mapping. All forms contain information on person–land relationships. Every case of crowdsourcing is based on a set of concepts that is used to record the environment. This is related to the knowledge that the group has. The category recordation vocabulary emerges, which can have very different forms depending on the context of information gathering. Existing vocabularies are often modified at the beginning of crowdsourcing projects by training programs (Flintan 2012; Ramirez-Gomez et al. 2013).

Reflections

Frank (2001) developed a five-tier ontology for geographical information, the tiers being *human-independent reality, observation of physical world, objects with properties, social reality,* and *subjective knowledge*. Each tier has its own rules for consistency. The vocabularies used in crowdsourcing cases do not normally fit into one specific tier, so consistency rules will probably be more complicated. Andersson and Chan (2014) state that many scholars have addressed the relationship between humans and the environment, and in particular how the different types of natural resources shape what humans may and may not do with them. The model for natural resource governance that they present (Andersson and Chan 2014, p. 18) gives strong relationships between natural resources and the rules that are governing their use, but the two are conceptualized differently. When studying participatory mapping it is therefore important to determine whether resources are mapped or the rules under which these resources are used.

State-Challenging Attitude

Many cases describe examples of land recordation in which the state is challenged or even absent. De Vries et al. (2014) qualify such systems as artifacts of dissatisfaction. Alternative land recordings are set up when a state-run land registration system exists but the information in it or part of it is not considered legitimate by certain social groups (nationally or internationally). Lengoiboni et al. (2011) describe the conflict between the mental recordings of pastoralists and the formally registered rights of non-pastoralist land use actors. In such cases, the building of sets of land recordation takes place under a different authority than the state. Such practices already have a long history. De Soto (2000, Chapter 5) describes the development of non-state land registration systems parallel to the official one in the United States in the nineteenth century.

Crowds and clouds in general are aware that the information they collect needs legitimacy to be accepted. The Map My Rights initiative (McLaren 2014) intends to build on the information that is supplied through crowd-sourced methods but introduces *trusted intermediaries* for quality control. The cases of bottom–up crowdsourcing that are described by De Vries et al. (2014) also show new non-state actors that have to approve the collected data. Archer et al. (2012) give such a role to *community architects*. Initiatives involving crowds and clouds are extensively published and in general backed by large numbers of national and international NGOs. In this way, they claim a high level of legitimacy.

In some cases, the claims of legitimacy are based on the model of the continuum of land rights (GLTN 2008, p. 8). This model supports "perceived tenure approaches" for which involvement of the state seems to be unnecessary. When land recording tools are based on such an approach, GLTN et al. (2012, p. 38) state that it is necessary to increase the acceptance of these tools by courts and land agencies. They describe cases in which this did not succeed.

The aforementioned examples lead to alternative recordation systems that can be easily distinguished from the state-run registration system. Practices such as corruption, discrimination, and land grabbing are also a challenge to the authority of the state if they are illegal from any legal perspective in the country in which they take place. Van der Molen and Tuladhar (2007) and Paresi et al. (2013) give many examples of corrupt practices. In such cases, the recordations based on these practices are much harder to distinguish from the registrations that are in conformity with the applicable law. Paresi et al. (2013) also present a toolkit to support transparency in land administration.

Reflections

Since land recordation takes place under a different authority than the state, the concept of participation as developed by Arnstein (1969) is not valid here, unless the concept is generalized for participation in decision making in any social context.

The state-challenging attitude is clear from the examples given in the former section. Such cases are in general well documented and explicit in their objective to seek recognition of rights and/or fight corruption or discrimination. Practices such as corruption and discrimination themselves can also be considered state challenging when they are against the law and especially when they are endemic. Corruption is generally recognized as an important factor to deal with in land registration (Van der Molen and Tuladhar 2007; Transparency International and UN-FAO 2012). However, concern about corruption is also expressed in relation to participatory methods of land registration (McLaren 2011). Also, discrimination can be executed by state as well as non-state actors. USAID (2004) and Todorovski et al. (2012) describe discriminatory practices at the state level in Kosovo between 1989 and 1999. In the literature on gender, many examples can be found of discrimination at the non-state level (Bicchieri and Knight 2014; Simbizi et al. 2014, p. 35).

Land grabbing and large-scale acquisition are terms that are used for the same phenomena. Wisborg (2013) defines land grabbing as "ethically unacceptable land acquisition." The large-scale acquisitor will refer to the deed of sale and the registration in the land register and call it a legal transaction. The acquisitor will claim that he or she found "empty land" (Borras et al. 2011). The reference for such an opinion is the operational registration law in the country because the state-run land registration system accepted the deed of sale. The question will be whether any right as defined by any of the entities of the deconstructed state is ignored by that transaction. If so, one can speak of land grabbing. Even if no national law is broken, non-state actors can still classify an acquisition as land grabbing using the framework of ethics of Wisborg (2013).

Holston (2008) describes the situation in Sao Paulo in which the authority of the state is challenged but the groups doing so are very skilled in using the legal instruments that are provided by the state. Holston states that in a certain context of time and place "land law promotes conflict, not resolution, because it sets the terms through which encroachments are reliably legalised." He qualifies this as insurgent citizenship. Askew et al. (2013) also describe many similar situations.

Categorizing

The selected literature deals with cases in which people make explicit their claims on land rights or the resources on which they depend. They specify the claims and resources, and they map the geographical boundaries of these claims and resources. In many cases, explicit specifications are not yet present in a form that is seen fit for the recordation process, so a classification of rights or resources is developed. Ramirez-Gomez et al. (2013) describe the design of a classification system for a participatory resource mapping project in the south of Surinam. Their dilemma was to find a method that both meets scientific standards and is understandable in the local language and culture. During a workshop, the villagers made a list of 90 landscape features that were important to them. These features were grouped into seven categories of landscape services. These services were regrouped into two categories of provisioning and cultural services following the Millennium Ecosystem Assessment typology (Millennium Ecosystem Assessment Program 2005). Based on the information from the villagers, two other categories were added: income services and touristic value. The four service groups were accepted as the mapping attributes (Laarakker et al. 2014). The case shows a final categorization of which the development is well documented. Barry and Meinzen-Dick (2008) describe a more general tenure box that can be used to classify rights. In other land recordation projects (Di Gessa 2008; Flintan 2012), the same process can be distinguished, but in several cases the design of the categories is not made explicit. Cloud-based initiatives have the same need for classification. One of the tasks that the Community Land Rights initiative sees for itself is to develop a typology of community lands (Community Land Rights 2014). Land Matrix (2013) uses a system that does not classify the types of rights but details the process steps necessary for a large-scale land transaction.

The described cases show an extensive use of the terms formal and informal. Crowdsourcing is often introduced as a means to formalize rights or to upgrade informal settlements.

Reflections

Categorization is bracketed out from the definitions of formal land registration quoted in the section "Theoretical Perspective." Land registration takes place in a situation in which a system to identify natural and legal persons and a system of land rights that can be registered are already in place. The role of the state in the registration process is limited to connecting identified persons to acknowledged rights on land. In the case studies, one can identify different state and non-state actors that are involved in the classification of land rights and tenure forms, classification of entitled persons and communities, and registration of the legal relations between these. The classification often takes place at the same time as the recordation.

Starr (1992) distinguishes between two meanings of categorization, on the one side the creation of classes and on the other side the assignment of a certain event in a class. Following Starr, the concept of categorization is only referring to the creation of classes of rights. In innovative approaches, the process of categorization (defining the categories) is often executed at the same time as the recordation (assignment of a certain man–land relation to a category). According to Starr (1992, p. 264), states have no choice but to categorize as follows:

> Every state must draw lines between kinds of people and types of events when it formulates criminal and civil laws, levies taxes, allocates benefits, regulates economic transactions, collects statistics, and sets rules for the design of insurance rates and formal selection criteria for jobs, contracts, and university admissions.

Starr considers such categories to be entrenched in the structure of institutions. The case studies show that non-state actors also have to categorize if they want to make their land claims clear. Also, these categories can get entrenched (Migdal 1988).

According to Barnes (2014), formal rules refer to legally binding agreements that are codified into policies, laws, regulations, or contracts. In most situations in which the concepts are used, formality stands for higher security, better protection, and registration by the state (Royston and Du Plessis 2014) and informality stands for the opposite. Labeling situations as formal or informal is an act of categorization.

Anderson (2002, p. 212) defines formal rights as the rights accepted by the state and distinguishes them from effective rights. This acceptance can take place at an international level, but Anderson finds that such rights are seen as fairly ineffectual expressions of principle:

> The international consensus of states on formal human rights is thus conceived in a different climate to that of formal legal rights within a state. Even autocrats tend to be more generous when agreeing in principle [...]. States certainly tend to behave with much greater caution about legislating domestic rights when their minds are more concentrated on their own affairs.

Anderson concludes that a separate social struggle may be required for converting formal rights into effective rights.

Starr (1992) describes how a political choice of category can become fixed as a cognitive commitment and component of social structure. A decision to frame a right, tenure system, or settlement as informal needs careful consideration. Wehrmann and Antonio (2011) point to the importance of the names of the rights that are awarded to people, independent of the exact legal meaning of that right. If one takes Anderson's approach, formality can be concluded much easier than when one uses the statutory land registration law as a reference. Enemark et al. (2014) plead to embed human rights in land administration systems. Human rights can also be used to qualify existing rights. Informal rights could be recategorized as formal (and legitimate) but not registered rights (UN-FAO 2012). If such labeling leads to upfront cognitive commitment, it could very well affect the added value of a state-run land registration project or a non-state-initiated crowdsourcing project. More research is necessary to understand this relationship.

Discussion

In this section, we bring the developed categories together, develop an enhanced conceptual framework, and discuss the result. The results described in the former section are based on the coding of publications that were selected based on an initial set of key words. The coding, category development, and checking of forward and backward citations lead to many more relevant key words and relevant publications that cannot all be studied in the framework of this chapter. The risk of too much complexity, which is mentioned by Verschuren and Doorewaard (1999) as a risk of the grounded theory methodology, is very valid. The methodology of Wolfswinkel et al. (2013) prescribes to revisit all previously coded literature if new search terms are developed. This has not been done, so the results discussed in this chapter describe the preliminary results of the grounded theory literature review. The developed conceptual framework is therefore presented in this section as a discussion.

The cases studies on crowds and clouds show a multitude of actors that are involved in the recordation of information on land rights, claims, and resources and the related processes of resolution of conflicts over them. Some of these actors openly confront the state, as can be seen from many case descriptions. Other actors exert their influence through more obscure processes. Power can be seen as the common denominator in all of this. Social power is necessary to be able to influence the recordation process on whatever social level it takes place. Power is necessary to make decisions on the categories of persons; categories of land; and categories of rights, claims, and resources that are recorded. Social power can be based on formal institutions that are applicable like the statutory legal framework or on informal institutions including endemic bad practices that are hard to prevent.

Any land recordation system has a center of power to which it refers. These centers take care of land recordation in the form of a statutory land registration system run by a state, the de facto tenure situation in a squatter area, the repeated stories told in a traditional community, and so on. The recordation is used by the center of power to rule on issues over land and the center has more power or less power to enforce this ruling, depending on the power of other social actors. If crowdsourcing takes place, these centers of power have to be taken into account since they can influence the end result.

These powers can be legitimate or illegitimate and state or non-state (Table 6.1). A more detailed analysis will reveal that both legitimacy and the level of functioning of state and non-state actors are points on a scale rather

TABLE 6.1
Powers

Power	Legitimate	Non-Legitimate
State	State authority	Corruption, land grabbing, discriminatory laws
Non-state	Non-state authority: religious, family, international, and so on	Gangs, warlords, exclusion, corruption

Source: Laarakker, D. M. et al., Person, Parcel, Power: An Extended Model for Land Registration + Powerpoint, XXV FIG Congress, Kuala Lumpur, Malaysia, 2014.

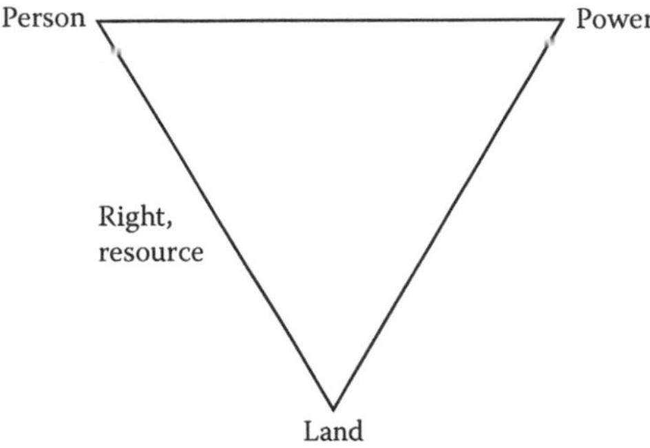

FIGURE 6.3
Power, person, and land.

than strict categories, but for the argument at this moment the dichotomies suffice. Legitimacy is a relative concept. Legitimacy is claimed based on legal or nonlegal arguments, and the social strength of the center of power will influence the acceptance of this claim by others. Legitimacy is a necessary concept because it gives the possibility to declare sets of information that are produced according to the applicable rules invalid. For this, a deeper analysis of the concept of law (Hart 1994) is necessary.

In Figure 6.3, the conceptual framework is further extended and the already deconstructed state in Figure 6.2 is replaced by the more general concept of power. The concept of power includes all possibilities of Table 6.1. The arrows in Figure 6.1 are left out because the relations are reciprocal; every power is also defined by the way it can control persons, land, and the relations between them. Different powers can have different views on a particular person–land relationship and may lay that view down in a land recordation system and base their protection mechanisms on it. Resource is added to the figure since this concept is often presented in a preliminary stage for claiming land rights. The model can possibly be further enhanced using the five tiers of ontology developed by Frank (2001). This needs further research.

The concept of power in the developed theoretical framework of Figure 6.3 emerged from the studied cases. It represents all acts that influence person–land relationships, both those in physical reality and those in administrative reality, and in the processes of law making, recordation, or conflict resolution. Power is seen as exercised by individuals or groups of individuals in smaller or bigger social structures that can influence the person–land relationship within a certain social setting.

Power is a concept that is extensively studied in scientific literature. According to Haugaard and Clegg (2009), the understanding of power is absolutely central to any understanding of society. Power is also a wide concept with many different meanings in different scientific contexts. It can be deconstructed in several ways. Balasz (2014) distinguishes social, sovereign, and governmental power. Social power stems from individual wills, sovereign power represents the collective will, and governmental power is based on the concept of public interest. Governmental power encloses judiciary, executive, and legislative powers. Balasz argues that

> once the collective agent [sovereign power] is meaningfully established, individuals turn into citizens while remaining individuals, too, and are prepared to make and work with the distinction between the mere collection of the wills of individuals and the necessarily unitary will of the collective person.

In cases of crowds and clouds, (state) government power is not accepted as being representative of the collective will (sovereign power). Therefore, alternative ways of meaningful establishment of the collective will are exercised and citizenship is not developed in the framework of the state, but in alternative social structures. This model of power seems to fit well to the cases of crowds and clouds, but further research is necessary.

Rye (2014a) distinguishes individualistic, strategic, bureaucratic, constitutive, and disciplinary power. All types are connected to certain elements of organization. This framework is primary developed to understand the exercise of power within political parties and also to understand political power more in general. Since land in any country is a very political issue, the framework seems very relevant. Rye (2014b) adapts his model further in his study on organizations with political and social goals. According to Rye, each mode of power represents different means by which participants in organizations with social and political goals may be either empowered as active citizens or prevented from being so. Such a model can be used to describe or predict the effectiveness of people in claiming their land rights (Holston 2008).

The literature in the domain of land administration also regularly refers to the concept of power. A large body of literature deals with issues of empowerment of the poor, women, marginalized groups, and so on. Barnes (2014) considers the distribution of power fundamental for addressing development problems such as malfunctioning land administration systems. Rambaldi et al.

(2006) warn about the power-changing effect of participatory mapping. Also, Williamson et al. (2010) regularly refer to the concept.

In the studied cases, power is also executed through categorization. In state-run land registration systems, statutory laws will define the types of rights and contracts on rights that can be registered. The history of such laws in its political context will reveal the powers that lead to a specific outcome of categorization. In case descriptions concerning crowds and clouds, the process of classification is described with different degrees of detail. Categorization is much about language; it is about naming resources and rights and giving meaning to those concepts. It is about translating mental recordings into written recordings. It is about translating social reality into administrative reality. Two decades ago, legal anthropology turned to linguistics-oriented research (Curran 2007) in addition to approaches that took rules or conflict resolution, as the starting point for their research. Further research is necessary to explain crowds and clouds also from this perspective. The legal and social meaning and effect of this categorization should be well understood, including that of formal or informal.

The continuum of land rights (GLTN 2008) and the concept of legitimate tenures in Voluntary Guidelines on the responsible governance of tenure (UN-FAO 2012) support the philosophy that non-state actors take care of the classification. This implies that the state could register rights that are not classified by the state but by non-state actors. However, one could argue that the acceptance by the state of classifications by non-state actors attributes state power to these formerly non-state actors.

Even though the concept of power is referred to in land administration literature regularly, it is not done in a structured way. An extensive analysis of the concept of power and its applicability to land recordation systems is too large for the context of this chapter, but it is a promising concept for further understanding.

Conclusion

Crowds and clouds linked to land registration emerge, in general, in a context of dissatisfaction with the state-run land registration system. A multitude of interrelated social actors are trying to protect their land interests by influencing the state-run land registration system or the alternative recordation systems that are set up by non-state actors. The influencing is done in a legal or an illegal way and in a legitimate or an illegitimate way. The end result of the recordation process and the related distribution of resources will depend on the distribution of power and authority between all these actors. Power and the related concepts of legitimacy and illegitimacy are therefore important to evaluate the innovative approaches introducing crowds and clouds.

When evaluating these innovative approaches, it is necessary to have an open mind. State-run land registration systems are not necessarily good or bad, nor are crowdsourced methods. A logical framework to evaluate and compare the results of both state-run systems and non-state-run systems should be able to explain the positive and negative effects of both approaches. The presented models can be seen as a first step in this development.

Clouds are relevant in the mobilization of public opinion about person–land relationships that are considered illegitimate in national and international discourses. As such, clouds are a source of power in the land debate, but it will depend very much on the authority that is given to them in the national context and whether they will be effective. Some of the web initiatives show acts of categorization, and it is important to follow these developments to be able to understand the long-term effects of this innovation.

References

Anderson, T. 2002. The political economy of human rights. *Journal of Australian Political Economy* 50:200–227.

Andersson, K., and J. Chan. 2014. Theory and conceptual foundations of natural resource governance. In *Adaptive Cross-Scalar Governance of Natural Resources*, edited by G. Barnes and B. Child. London, United Kingdom: Earthscan.

Archer, D., C. Luansang, and S. Boonmahathanakorn. 2012. Facilitating community mapping and planning for citywide upgrading: the role of community architects. *Environment and Urbanization* 24 (1):115–129. doi: 10.1177/0956247812437132.

Arnstein, S.R. 1969. Ladder of citizen participation. *Journal of the American Institute of Planners* 35 (4):216–224. doi: 10.1080/01944366908977225.

Askew, K., F. Maganga, and R. Odgaard. 2013. Of land and legitimacy: a tale of two lawsuits. *Africa* 83 (1):120–141. doi: 10.1017/s0001972012000745.

Bakker, L. 2006. Land and authority: the state and the village in Pasir, East Kalimantan. *IIAS Newsletter* 40:15.

Balasz, Z. 2014. *The Principle of Separation of Powers: A Defence*. World Congress of the International Political Science Association 2014, Montreal, Canada.

Barnes, G. 2014. Introduction in adaptive cross-scalar governance of natural resources. In *Adaptive Cross-Scalar Governance of Natural Resources*, edited by G. Barnes and B. Child. London, United Kingdom: Earthscan.

Barros, D.B. 2010. *Hernando de Soto and Property in a Market Economy, Law, Property and Society*. Farnham, England; Burlington, VT: Ashgate.

Barry, D., and R. Meinzen-Dick. 2008. *The Invisible Map: Community Tenure Rights*. Paper presented at the IASC Conference: Governing Shared Resources: Connecting Local Experience to Global Challenges, Cheltenham, United Kingdom, July 14–18.

Barry, M., and K. Asiedu. 2014. *Tracking Changing Tenure Relationships: The Talking Titler Methodology Using Data Mining and Social Network Analysis*. World Bank Conference on Land and Poverty: Integrating Land Governance into the Post-2015 Agenda, Washington, DC.

Bennett, R.M. 2007. *Property Rights, Restrictions and Responsibilities: Their Nature, Design and Management*. Parkville, Australia: University of Melbourne.

Bennett, R.M., P. van der Molen, and J.A. Zevenbergen. 2012. Fitted, green, and volunteered, legal and survey complexities of future boundary systems. *Geomatica* 66 (3):181–193.

Bicchieri, M., and R. Knight. 2014. *It Takes a Village to Protect a Woman's Land: Working at the Community-Level to Secure Woman's Land Rights in Southern Africa*. World Bank Conference on Land and Poverty: Integrating Land Governance into the Post-2015 Agenda, Washington, DC.

Borras, S.M., R. Hall, I. Scoones, B. White, and W. Wolford. 2011. Towards a better understanding of global land grabbing: an editorial introduction. *Journal of Peasant Studies* 38 (2):209–216. doi: 10.1080/03066150.2011.559005.

Community Land Rights. 2014. Global Map of Indigenous and Local Community Lands, accessed October 13, 2014, http://www.communitylandrights.org/wp-content/uploads/2014/05/Global-Map-2-pager.pdf.

Curran, J. 2007. States of the Art: A History of Legal Anthropology and the Next Generation of Research at the Intersection of Language, Ideology, and Power, accessed December 02, 2014, http://johncurran.files.wordpress.com/2007/11/john-curran-a-history-of-legal-anthropology-and-the-next-generation-of-research-final-draft.pdf.

De Soto, H. 2000. *The Mystery of Capital: Why Capitalism Triumphs in the West and Fails Everywhere Else*. New York: Basic Books.

De Vries, W.T., R.M. Bennett, and J.A. Zevenbergen. 2014. Neo-cadastres: innovative solution for land users without state based land rights, or just reflections of institutional isomorphism? *Survey Review*, in press.

De Vries, W.T., and J. Zevenbergen. 2011. Discretionary space as a concept to review innovation in land administration in Africa. *Survey Review* 43 (323):638–652. doi: 10.1179/003962611x13117748892399.

Di Gessa, S. 2008. Participatory mapping as a tool for empowerment. In *ILC Knowledge for Change Series*. Rome, Italy: International Land Coalition.

Enemark, S., L. Hvingel, and D. Galland. 2014. Land administration, planning and human rights. *Planning Theory* 13 (4):331–348. doi: 10.1177/1473095213517882.

Frank, A.U. 2001. Tiers of ontology and consistency constraints in geographical information systems. *International Journal of Geographical Information Science* 15 (7):667–678. doi: 10.1080/13658810110061144.

Georgiadou, Y., J.H. Lungo, and C. Richter. 2014. Citizen sensors or extreme publics? Transparency and accountability interventions on the mobile geoweb. *International Journal of Digital Earth* 7 (7):516–533. doi: 10.1080/17538947.2013.782073.

Gilbert, A. 2002. On the mystery of capital and the myths of Hernando de Soto: what difference does legal title make? *International Development Planning Review* 24 (1):1–19. doi: 10.3828/idpr.24.1.1.

GLTN. 2008. *Secure Land Rights for All*. Nairobi, Kenya: GLTN/UN-Habitat.

GLTN, UN-Habitat, and IIRR. 2012. *Handling Land: Innovative Tools for Land Governance and Secure Tenure*. Nairobi, Kenya: UN-Habitat, IIRR.

Goodchild, M.F. 2009. Neogeography and the nature of geographic expertise. *Journal of Location Based Services* 3:82–96.

Goodchild, M.F., and J.A. Glennon. 2010. Crowdsourcing geographic information for disaster response: a research frontier. *International Journal of Digital Earth* 3 (3):231–241. doi: 10.1080/17538941003759255.

Graeber, D. 2007. Provisional Autonomous Zone, or the Ghost-State in Madagascar. In *Possibilities: Essays on Hierarchy, Rebellion, and Desire*, 155–180. Oakland, CA: AK Press.

Griffith-Charles, C. 2011. The application of the social tenure domain model (STDM) to family land in Trinidad and Tobago. *Land Use Policy* 28 (3):514–522. doi: 10.1016/j.landusepol.2010.10.004.

Haklay, M., V. Antoniou, S. Basiouka, R. Soden, and P. Mooney. 2014. *Volunteered Geographic Information in Government*. Report to GFDRR (World Bank). London, United Kingdom: World Bank.

Hart, H.L.A. 1994. *The Concept of Law*. 2nd ed. Oxford, NY: Clarendon Press; Oxford University Press.

Harvey, F. 2006. Elasticity between the cadastre and land tenure: balancing civil and political society interests in Poland. *Information Technology for Development* 12 (4):291–310. doi: 10.1002/itdj.20049.

Haugaard, M., and S.R. Clegg. 2009. Why power is the central concept of the social sciences. In *The Sage Handbook of Power*. Thousand Oaks, CA: Sage Publications.

Henssen, J.L.G. 1995. Basic principles of the main cadastral systems in the world. In *Modern Cadastres and Cadastral Innovations, Proceedings of the One Day Seminar in Delft* 5–12. Delft, The Netherlands: FIG Commission 7 and University of Melbourne.

Holston, J. 2008. *Insurgent Citizenship. Disjunctions of Democracy and Modernity in Brazil*. Princeton, NJ: Princeton University Press.

Flintan, F. 2012. *Participatory Rangeland Resource Mapping As a Valuable Tool for Village Land Use Planning in Tanzania*. Rome, Italy: International Land Coalition.

Karsten, N., and L. Timmers. 2008. To read or not to read: over de waarde van vakliteratuur in kwalitatief onderzoek. *Kwalon* 39 13(3):5–11.

Laarakker, P.M., and W.T. De Vries. 2011. *Exploring Potential Avenues and Concerns*. www.opencadastre.org. FIG Working Week 2011, Marrakech, Morocco.

Laarakker, P.M., J.A. Zevenbergen, and P.Y. Georgiadou. 2014. *Person, Parcel, Power: An Extended Model for Land Registration + Powerpoint*. XXV FIG Congress, Kuala Lumpur, Malaysia.

Land Matrix. 2013. Newsletter June 2013, accessed October 13, 2014, http://landmatrix.org/media/filer_public/2013/06/10/lm_newsletter_june_2013.pdf.

Larsson, G. 1991. *Land Registration and Cadastral Systems. Tools for Land Information and Management*. Harlow, United Kingdom: New York: Wiley; Longman Scientific and Technical.

Lemmen, C.H.J. 2012. *A Domain Model for Land Administration* PhD, Delft University of Technology, NCG publications on Geodesy no. 78. Netherlands Geodetic Commission.

Lemmen, C.H.J., C. Augustinus, P.J.M. van Oosterom, and P. van der Molen. 2007. *Social Tenure Domain Model: Design of a First Draft Model*. FIG Working Week 2007, Hong Kong SAR, China.

Lemmen, C.H.J., and D. Haarsma. 2012. Rwanda brings 10 million parcels under registered title: GIM international interviews Emmanuel Nkurunziza, director general of the Rwanda Natural Resources Authority. *GIM International* 26 (6):14–19.

Lengoiboni, M., A.K. Bregt, and P. van der Molen. 2011. Pastoralists Seasonal Land Rights in Land Administration: A Study of Northern Kenya. 183, ITC diss., Wageningen University, The Netherlands.

Li, T.M. 2005. Beyond "the state" and failed schemes. *American Anthropologist* 107:383–394.

McCall, M.K. 2011. *The Power of Participatory Mapping for Mobilising Indigenous Knowledge on Property Rights*. Pastoralist Seasonal Land Rights in Land Administration. Enschede, The Netherlands: ITC, University of Twente.

McCall, M.K., and C.E. Dunn. 2012. Geo-information tools for participatory spatial planning: fullfilling the criteria for good governance. *Geoforum* 43 (1):81–94.

McLaren, R. 2011. *Crowdsourcing Support of Land Administration. A New, Collaborative Partnership between Citizens and Land Professionals*. RICS Research Report, World Bank Conference on Land and Poverty, Washington, DC.

McLaren, R. 2014. *Can the MapMyRights Initiative Be a Game Changer*? World Bank Conference on Land and Poverty: Integrating Land Governance into the Post-2015 Agenda, Washington, DC.

McLaughlin, J., and S. Nichols. 1989. Resource management: the land administration and cadastral systems component. *Surveying and Mapping* 2:77–86.

Migdal, J.S. 1988. *Strong Societies and Weak States: State-Society Relations and State Capabilities in the Third World*. Princeton, NJ: Princeton University Press.

Millennium Ecosystem Assessment Program. 2005. *Ecosystems and Human Well-Being: Synthesis*. Washington, DC: Island Press.

Nichols, S., and J. McLaughlin. 1990. *The Information Role of Land Registration in Land Administration*. FIG XIX International Congress, Helsinki, Finland.

Ozanne, J.L. 1992. Book review: basics of qualitative research by A. Strauss and J. Corbin. *Journal of Marketing Research* 29 (3):382–384. doi: 10.2307/3172751.

Paresi, C., S. Haile, M. Permezel, S. Asiama, W. Kombe, J. Gold et al. 2013. *Tools to Support Transparency in Land Administration: Securing Land and Property Rights for All: Training Package: Trainers' Guide*. Nairobi, Kenya: GLTN, UN-HABITAT.

Patel, S., and C. Baptist. 2012. Documenting by the undocumented. *Environment and Urbanization* 24 (1):3–12.

Rambaldi, G., R. Chambers, M. McCall, and J. Fox. 2006. Practical ethics for PGIS practitioners, facilitators, technology intermediaries and researchers. *Participatory Learning and Action* 54:106–113.

Ramirez-Gomez, S., G. Brown, and A. Tjon Sie Fat. 2013. Participatory mapping with indigenous communities for conservation: challenges and lessons learned from Surinam. *Electronic Journal on Information Systems in Developing Countries* 2:1–22.

Richter, C., and P.Y. Georgiadou. 2014. Practices of legibility making in Indian cities: property mapping through geographic information systems and slum listing in government schemes. *Information Technology for Development*, in press.

Royston, L., and J. Du Plessis. 2014. *A Continuum of Land Rights: Evidence from southern Africa*. World Bank Conference on Land and Poverty: Integrating Land Governance into the Post-2015 Agenda, Washington, DC.

Rye, D. 2014a. *Political Parties and the Concept of Power, a Theoretical Framework*. London, United Kingdom: Palgrave Macmillan.

Rye, D. 2014b. *The Concept of Power in the Analysis of Organisations with Social and Political Goals*. World Congress of the International Political Science Association 2014, Montreal, Canada.

Scott, J. 1998. *Seeing like a State—How Certain Schemes to Improve the Human Condition Have Failed*. New Haven, CT: Yale University Press.

Seventh International Conference of American States. 1933. Convention on Rights and Duties of States. Montevideo, Uruguay.

Simbizi, M.C.D., R.M. Bennett, and J. Zevenbergen. 2014. Land tenure security: revisiting and refining the concept for Sub-Saharan Africa's rural poor. *Land Use Policy* 36:231–238. doi: 10.1016/j.landusepol.2013.08.006.

Singh, A. 2014. *Incremental Tenures and Service Delivery in Low Income Irregular Settlements in Indian Cities*. World Bank Conference on Land and Poverty: Integrating Land Governance into the Post-2015 Agenda, Washington, DC.

Starr, P. 1992. Social Categories and Claims in the Liberal State. *Social Research* 59 (2):263–295 (Summer 1992).

Strauss, A.L., and J.M. Corbin. 1990. *Basics of Qualitative Research: Grounded Theory Procedures and Techniques*. Newbury Park, CA: Sage Publications.

Strauss, A.L., and J.M. Corbin. 1998. *Basics of Qualitative Research: Techniques and Procedures for Developing Grounded Theory*. 2nd ed. Thousand Oaks, CA: Sage Publications.

Todorovski, D., J.A. Zevenbergen, and P. Van der Molen. 2012. Land administration in post-conflict environment—aspects relevant for South East Europe. *South-Eastern European Journal of Earth Observation and Geomatics* 1 (2S):47–59.

Transparency International and UN-FAO. 2012. *Corruption in the Land Sector*. Berlin, Germany.

UN-FAO. 2012. *Voluntary Guidelines on the Responsible Governance of Tenure of Land, Fisheries and Forests in the Context of National Food Security*. Rome, Italy: UN-FAO.

USAID. 2004. *An Assessment of Property Rights in Kosovo*. Final report, March. Prishtina, Kosovo: USAID.

Van der Molen, P., and A.M. Tuladhar. 2007. Transparency in land administration: corruption is everywhere. *GeoInformatics* 10 (4):12–15.

Van Egmond, N.D., and H.J.M. de Vries. 2011. Sustainability: the search for the integral worldview. *Futures* 43 (8):853–867. doi: 10.1016/j.futures.2011.05.027.

Van Leeuwen, M. 2014. Renegotiating customary tenure reform—land governance reform and tenure security in Uganda. *Land Use Policy* 39:292–300. doi: 10.1016/j.landusepol.2014.02.007.

Vaquero, L.M., L. Rodero-Merino, J. Caceres, and M. Lindner. 2009. A break in the clouds: towards a cloud definition. *Acm Sigcomm Computer Communication Review* 39 (1):50–55.

Verschuren, P., and H. Doorewaard. 1999. *Designing a Research Project*. Utrecht, The Netherlands: Lemma.

Von Benda-Beckmann, F. 2003.Mysteries of capital or mystification of legal property? *Focaal—European Journal of Anthropology* 41:187–191.

Wehrmann, B., and D. Antonio. 2011. Intermediate land tenure: inferior instruments for second-class citizens? *FAO's Land Tenure Journal* 2011(1):6–25.

Williamson, I.P, S. Enemark, J. Wallace, and A. Rajabifard. 2010. *Land Administration for Sustainable Development*. Redlands, CA: ESRI Press Academic.

Wisborg, P. 2013. Human rights against land grabbing? A reflection on norms, policies, and power. *Journal of Agricultural and Environmental Ethics* 26 (6):1199–1222. doi: 10.1007/s10806-013-9449-8.

Wolfswinkel, J.F., E. Furtmueller, and C.P.M. Wilderom. 2013. Using grounded theory as a method for rigorously reviewing literature. *European Journal of Information Systems* 22 (1):45–55. doi: 10.1057/ejis.2011.51.

Zevenbergen, J.A. 2002. *Systems of Land Registration: Aspects and Effects*. Netherlands Geodetic Commission NCG: Publications on Geodesy no. 51, Netherlands Geodetic Commission (NCG), Delft, Netherlands.

Zevenbergen, J., C. Augustinus, D. Antonio, and R. Bennett. 2013. Pro-poor land administration: principles for recording the land rights of the underrepresented. *Land Use Policy* 31:595–604. doi: 10.1016/j.landusepol.2012.09.005.

Section III

Creating Innovative Designs

7

Point Cadastre

Walter T. de Vries, Co Meijer, Susan Keuber, and Bert Raidt

CONTENTS

Introduction

Most land information systems rely on systematic records of property rights that are associated with parcels and parcel boundaries (Tuladhar et al. 2004). Parcels may however not always be the most appropriate spatial basis for maintaining land records. Not only does surveying parcels and parcel boundaries tend to be too expensive in low-income countries (de Vries et al. 2003), but it also tends to be bureaucratic in countries with weak governance systems (Hanstad 1997) and deferred when limited technical surveying resources are available (Enemark and Williamson 2004). In such cases, alternative organizational processes are required as a basis for developing and maintaining the land records. One such alternative is the point cadastre.

A point cadastre is stripped-down method for collecting and maintaining cadastral data (Hackman-Antwi et al. 2013). Geographic points, instead of boundaries or areas, are used as the key reference to represent land parcels in a cadastral database. One of the main requirements to recognize and distinguish these points is that they should be uniquely identified (Fourie 1994), so that they can be connected to tenure attributes, such as owner names or land rights. The concept *point cadastre* has many equivalents in literature, such as center-point cadastre (Fourie 1994), lots-by-dots (Burke 1995), single-point and midpoint cadastres (Home and Jackson 1997), or dots on plots (Davies and Fourie 2002). Experiments of point cadastres (lots-by-dots) designs date back to early 1990s to document rural parcels in Honduras,

Indonesia, Pakistan, and the Philippines (Burke 1995) and of recording land rights in slum areas in South Africa (Home and Jackson 1997). The name suggests that technically one does not bother with the size, extent, or boundary of an area with homogeneous rights, yet relates the land information to a single surveyed or identified point—either centroid or midpoint. We use the term "point cadastre" for the remainder of this article.

Although the reference to points in the approach suggests that the added value refers to the physical shape of geospatial feature in the database, Keuber (2014) argues that the socio-institutional situation of land tenure is an intrinsic aspect of point cadastres as well. The point as such represents more than the *dot*. It represents a relation with rights and claims and offers a means to both recognize tenure and social relations, which would otherwise be hidden from administration. An additional advantage of point cadastres as compared to conventional cadastres concerns the potential of savings in cost and other resources. The cost savings are especially in the process of constructing and populating the spatial database. Instead of having to rely on individual land surveys and adjudication processes constructing the boundaries to be included in the databases, one populates the database with a point as basic geometric reference and connects other attributes to such a point. Hence, when searching for low-cost approaches to data acquisition in land administration, point cadastre can be seen as a very usable starting tool.

Conventional cadastral systems rely heavily on expert practitioners, extensive quality procedures, and sustained financial resources. The underlying assumption hereby is that such capacity is available and can directly be put to use to construct and maintain a cadastral database. Such an approach may of course often be the preferred option for governments who aim for the best quality—both in geometry and in content. However, if a government (including local governments) aims to improve or start-up their land administration system, and thereby rely on standards and rules set in Western countries, then it may take a long time before the system can actually be used. The professional stakeholders, such as surveyors and registrars who would act as basic actors to populate the database, would need to have the skills to do so and should also be available in sufficient numbers. Experience has shown that developing such capacity may however take various decades, which is not the time frame that governments have in mind. Consequently, alternative systems that can be used immediately, while being developed much quicker and being maintained with relative limited capacities and resources, are preferable in such cases.

Many countries need to improve their management of the land. What is described in this chapter can be a very good starting point to set up quick and simple first registrations. In this way, the system can be improved step by step later. This chapter deals with how to build a point cadastre, whereby it identifies the challenges needed to be overcome. These include both technical, organizational, management, and institutional challenges, as it is important that the design does not occur in isolation of the context but is done in a responsible way. It is shown that a point cadastre has to be clearly based on an identified need (to have a cadastre and cadastral info), a request by stakeholders and a validation process of discussions and feedback of stakeholders.

Theoretical Perspective

Conceptually, the point cadastre approach has both a historical and a geospatial or information technical root. Historically, it has especially been considered suitable and appropriate in locations where weak interests in land prevail, where informal tenure is not yet aligned with a formal registration system, or where connections to other registers are lacking (Deininger et al. 2010). This may in particular be true where there are gaps between the recognition of formal (registered) and informal (unregistered) tenure, or in areas of (post) conflict. Point cadastre may then be an alternative method based on relatively rapid data collection and registration.

From a geospatial and information technical perspective, the point cadastre is a cadastral system that uses points instead of parcels, or otherwise put, uses points to represent closed areas (either parcels or buildings). Point instead of parcel identifiers thus becomes the key. Technically, a point cadastre can serve as a multipurpose cadastre or land information infrastructure. As a basic register it can also be used as a starting point for a street name and house number register. It thus has the potential to be linked to other registers using these identifiers. It then serves a broader range of domains: not only to provide tenure security but also for basic management issues and can be used by local or central government, by health services, and by other service deliverers, such as water, electricity, as an instrument to support valuation and taxation and the execution of a census. Even transactions on objects related to the point can be handled.

Point cadastre may serve a different purpose and be designed differently for urban or rural areas. Furthermore, the practice of record keeping may be different in rural versus urban point cadastres. Although the records associated with point cadastres should provide easier access to land and security of tenure than before, in a rural setting the record related to the land is most crucial whereas in urban setting the record of the house/building providing the location of shelter may be more crucial. This has an influence on the design of the underlying (spatial) database model.

An appropriate reference to evaluate a design of a point cadastre is the framework of requirements engineering (van Vliet 2008). This framework starts from the assumption that technology cannot function in isolation from its environment and vice versa. A technical design therefore not only goes beyond translating technical requirements into technical possibilities, but also recognizes social and cognitive tacit knowledge in the design and implementation practice. Such tacit knowledge is often associated to experience with what works and what does not in a given local physical and institutional context. In short, the requirements analysis engineering includes three major aspects:

1. Evaluation of technical requirements. For point cadastres, Hackman-Antwi (2012) and Hackman-Antwi et al. (2013) differentiate three categories of main requirements for point cadastres: functional requirements (e.g., related to establishment of titles, boundaries or legal records, to maintenance and interoperability functions), quality requirements (related to ease of use, cost, flexibility, scalability, and accuracy), and architectural requirements (related to data collection, storage, maintenance editing tools, visualization, field infrastructure).
2. Adaptation and incorporation of physical conditions of the area. This primarily relates to the degree of stability and access to the area for which a point cadastre is made is required. In case of landslides or floods, for example, the physical shape may be subject to such changes that physical structures may no longer be recognizable and people may be forced to leave the area. In addition, when parcel boundaries already coincide with physical boundaries, representation of a point instead of an area may not provide any additional advantage.
3. Legal framework and stakeholder environment in which the new cadastre needs to be embedded. This relates to the degree to which a new type of cadastre can either be embedded in currently operational legislation or whether a new law or regulation needs to be made. The former strategy is easier to adopt but requires careful examination in legislation whether technical changes are legally valid. The latter requires introduction of new legislation and is thus more difficult to adopt. Yet, it would have the advantage that the administration of the law could be constructed directly alongside the technical system.

The collection of these requirements provides the framework for practitioners designing solutions of point cadastres. This theoretical framework needs however testing in a practical setting. This chapter does this.

Methodology

The requirements analysis engineering framework forms the basis to evaluate each of the designs of point cadastres in two specific cases. The comparison of the two cases is done qualitatively, with the aim to derive a set of a generic recommendation how and when point cadastres could be designed and used. The base data from the two cases partly draw on secondary sources and partly on direct communication and personal experience. Data are drawn from recent work on developing a point cadastre for Bugala Island in Uganda (Keuber 2014) and in Guinea-Bissau based on experiences of Kadaster International. Bugala Island is in the south of Uganda in Lake Victoria. It has approximately 10,000 ha of land developed for a vegetable oil palm plantation, out of which 3,500 ha are for small-hold farmers and 6,500 ha for the nuclear farm itself. It is a suitable location for developing a point cadastre given that smallholder farmers' land is not documented. Most of the smallholders are tenants on the land, and although the law provides the possibility to issue certificates of occupancy and guarantee some sort of tenure security, the landlords are reluctant to do so. A point cadastre can help in establishing a basic record of tenancy.

Guinea-Bissau is a country in Western Africa. The mayor and the council of the capital Bissau had challenges in managing their city. There was a lack of information to manage the city, because of a bad quality of paper-based administration, low capacity, computer illiteracy, and outdated registers based on several Microsoft (MS) Access databases. These were however not considered reliable any more. Furthermore, there was a lack of topographical maps on the basis of which planning and administration could be carried out. During a fact finding mission in 2011 on request of the Mayor of the city, the advice and first outline was given to establish a low-cost and simple-to-use registration, with a direct link to a geospatial reference. This paved the way to design an innovative multipurpose point cadastre.

The requirement specifications for Bugala Island were drawn from a collection of sources, including the technical point cadastre requirements as specified by Hackman-Antwi (2012), whereas the data on Bugala physical conditions and the Bugala stakeholders were drawn from the empirical work by Keuber (2014). She conducted interviews with stakeholders in the area and reported on needs and requirements, and also drew conclusions on how best to design an associated land information system. The development of the prototype relied on 9 personal interviews and 17 survey responses (Keuber 2014), and data modeling based on unified modeling language (UML) and land administration domain model (LADM) specifications, all supplemented by the architectural requirement for the underlying information and communication technology (ICT). The prototype was tested in a pilot area, Kasekulo village, with 57 parcels (represented by points). In addition, a thorough stakeholder requirement and feedback analysis was executed to capture social and cognitive aspects. Additional documentary evidence from Uganda and its legislation on land matters relied on formal and gray literature relating to the Ugandan Land Act of 1998.

The development of the Guinea-Bissau prototype relied on a fact-finding mission on invitation by the Camara Municipal de (City Council of) Bissau in July 2011 and subsequent discussions on design and functionality at Kadaster International in August/September 2011. The city needed advice on how to improve its administration. In Bissau field visits, meetings with municipal and governmental stakeholders, the tax department, and some ministries gave insight in the existing situation. Meetings with the mayor gave first insight in the questions and desired functionalities that were needed to start and improve the administration and management of the city. A workshop was organized to present and discuss the first ideas and advises with a broad audience from local and central governments.

Results

In the Bugala Island case, the exploratory interviews with stakeholders identified several basic current problems of tenants (Keuber 2014), including perceived bureaucracy when having to hire surveyors privately and communicating with the Ministry of Lands, no issuance of land tenure documentation, no available or accessible expertise or authority in handling land conflicts, and dependence on the goodwill of local political leaders. In addition, the regional office in Masaka indicated having problems with both squatters and absentee landlords, lack of occupancy information leading to incomplete local register, and hesitance to record rights at all. A point cadastre had to address these problems alongside establishing a new system of recording rights. Given these observed problems preparatory requirements were formulated before addressing the technical and legal requirements and crafting the prototype. Such preparatory requirements included the establishment of a permanent body that would eventually operate and maintain the point cadastre, attracting and training skilled personnel to operate the system and ensuring that the output of the new system would receive legal status (being a legal document related to land). It was agreed that with these requirements in place there would be a suitable ground for establishing the point cadastre. The strategy to do so addressed the specific elements of the requirements engineering methodology. Each of the elements was discussed in consultation with the stakeholders.

Technically, the stakeholders indicated that any geospatial information with the point cadastre should be georeferenced using GPS coordinates. Any data collection on points should be done relatively quickly to make a significant difference in comparison to conventional methods. Parcel boundaries were considered relevant for tenants, but they were satisfied with the option to link the points of a point cadastre to the parcel sketches. Photographs of owners and tenants should be included whenever possible. Finally, scalability of the system was considered crucial. Connections to other databases, both spatial and nonspatial, should be easy. Access to Internet server and mobile telephone networks would even increase this possibility.

The combination of these requirements resulted in the following preliminary technical design for the point cadastre in Bugala Island (Figure 7.1). The model is adapted yet based on the international standards captured in the LADM. The class of occupied parcel, which describes the geometry of the parcel with a point, can at the same time also be connecting sketch plans reflecting an area. Tenants are furthermore identified by pictures among others, which would make it easier and more transparent when recalling with whom the arrangement was made.

The physical conditions of the area were such that the area was relatively flat with good access to parcels. This is a crucial characteristic when relying on low-end global navigation satellite system (GNSS) receivers to acquire single, georeferenced points and aiming to make sketch plans by hand.

As far as using and/or adapting the legal framework is concerned, first observations showed that very limited occupancy was supported in some sort of register, which would be acknowledged by the current legal framework. Given that the only legal instruments to provide this framework were

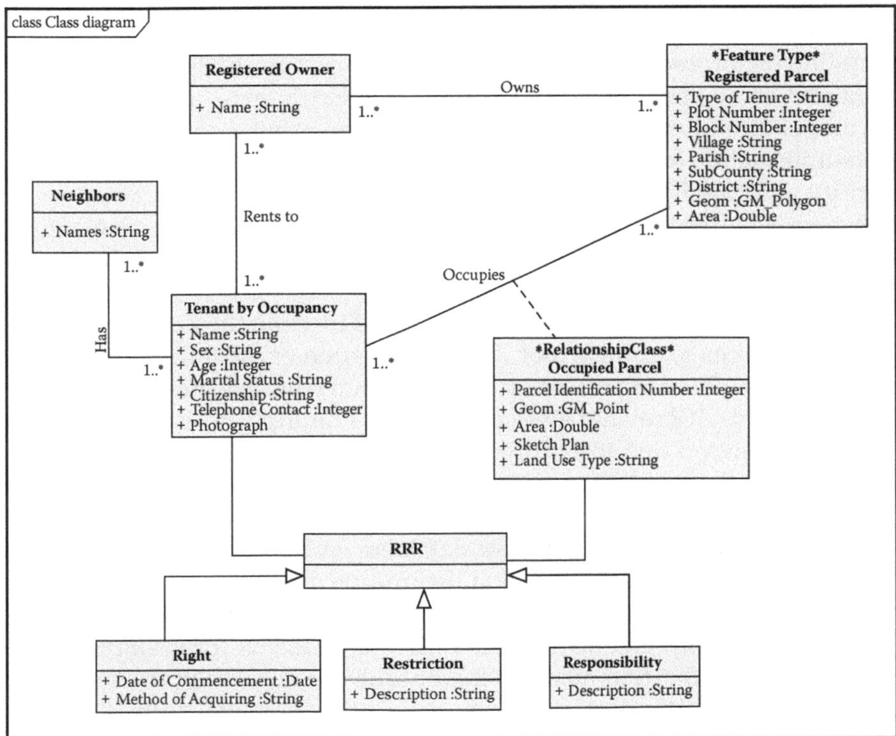

FIGURE 7.1
Preliminary conceptual data model of Bugala Island.

the land act of 1998 and its related regulations of 2004; any operational adjudication process leading to points should fit in these two legal instruments. Consequently, stakeholders were consulted about which information they would like to see recorded in the new point cadastre system, and these needs were compared and translated into the specific attributes described in the land act and the regulations. This included for example the determination of "who owns what and where," but also the spatial–legal relation between major landlords and land occupants needed to be recorded during an adjudication and demarcation process. Once the prototype containing the point cadastre records was completed all farmers were requested to validate the information that could be traced to them.

In the Guinea-Bissau case, the functional requirements were acquired through a visit and meeting with the stakeholders in Guinea-Bissau, after which the development of the Guinea Bissau prototype relied on a brainstorm session with land administration experts and geographic information system (GIS) experts and software developers in 2011. This session was organized to check and discuss and even pilot the ideas to find out if point cadastre and the management of it would be viable. This resulted even in a rough first pilot version that was tested and made available in the cloud in a cooperative action of the Netherlands Kadaster and GIS software giant ESRI.

The technical implementation of the prototype was executed by proposing the following procedure: each object (physical structure, house, etc.) had to be identified on an aerial photo and/or a Google Earth map—usually through the roof of the structure. These aerial photos were printed and provided with meta-information (name of city, neighborhood, and number of the photo) to identify and classify each photo. Inspectors—in the form of para surveyors and para legal practitioners—were identified to classify each element on the photos. This process was demonstrated through several examples, as depicted in Figure 7.2.

The procedure continued by identification of roofs on the ground and connection of points to a single roof. The combination of the unique number of the dot on the photo (representing a roof) with the unique identifier number of the photo itself provides a basic record of a multipurpose cadastre. This can also generate a link to existing administration of buildings, by connecting the dot to the building, or connecting the X and Y coordinates of the dot to that of the building. If there can be used a GIS system for the georeferencing and a digital administrative database on buildings, the connection between the administrative data and the unique dot number will have to be managed in the system. The above activities deliver a GIS database with the location of the dots, and an administrative database on for example buildings with the connection made by the unique dot number of the roof. This approach also makes it possible to register other essential sectors such as health, water, electricity, valuation, taxes, and real estate.

FIGURE 7.2
Prototype of point cadastre in Guinea-Bissau.

The physical conditions of the area are specific. The capital city of Bissau is a relatively flat and small area of approximately 10 × 10 km². Yet, the physical structures are closely and densely connected and usually contain only one floor. On the ground, people do not perceive clear boundaries of plots by physical structures or signs nor by verbal agreements. This physical situation and perception of stakeholders made the introduction of a point cadastre easier and more appropriate than having to measure individual boundaries.

With regard to the legal embedding of the new point cadastre stakeholders in Bissau explicitly called for the design and immediate implementation of a fit-for-purpose cadastre during the preparation phase. The specifications for Guinea-Bissau were collected during observations, discussions, and interviews with stakeholders in local and central government that had to deal with urban and rural migration, health issues, uncontrolled city development, land disputes, and underdeveloped valuation and taxation. The basis of these observations gave input to a first outline of what is realistically possible to develop in a step-by-step approach. From this the plans could be drawn for first steps in improving the land management functions within the government.

A brainstorm meeting was organized in the Netherlands to discuss the Bissau case. Discussions with IT experts and other advisors resulted in a first outline for solution to support a multipurpose point cadastre. Unfortunately, the actual construction of the point cadastre in Bissau could never materialize. Soon after the consultation and design process political unrest in the country and a failed coup attempt led to changes in the administration. Because of these circumstances, the process was immediately interrupted and the use of the system could never be tested.

Discussion

The experiences in the two cases provide insight in conditions under which a point cadastre may be useful, the technical model with which a point cadastre can be constructed, and the procedure that can populate and update point cadastre.

First of all, the physical conditions of the location requiring some sort of registration are significant when aiming to introduce a point cadastre. Although it is always possible to identify points on an aerial or satellite image, locating, recognizing, and connecting these points on the ground may be seriously hampered in rough areas and in areas with many overlapping buildings or structures. Still, both cases have shown that points were easily recognizable by stakeholders familiar with the location. Given that this issue came up in both cases it is valid to state that this condition applies for both rural and urban areas.

Second, both the role of stakeholder consultation and demand-driven approaches proved valuable. The tests reveal that the validation of requirements with local communities provides a better understanding among stakeholders of why the introduction of a new type of cadastre could be beneficial. It also acts as a vehicle to open up tenure and governance discussions. Connecting furthermore, the construction of the system as a way to overcome concrete organizational constraints of a ministry or of a local government in a given context is paramount: it forces the system designers to connect functional processes to societal roles and benefits of local communities.

Third, on the technical requirements the pilot of Guinea-Bissau showed that in the field para surveyors can execute their activities without any support by equipment. A pen and an aerial photo on print are sufficient for data gathering or updating. There is no need for technical tools that need technical skills, batteries, digital connection with the GIS system, maintenance and costs, and so on. In the office on a GIS system, the collected dots on the roofs can be georeferenced on the aerial photo in the GIS database. Little technical requirements are needed, and in this low cost and with less (technical) capacity the results can be achieved. The georeferenced aerial photo in the GIS system should be updated as much as possible to have the most actual situation on the screen. This paves the way for active maintenance of the database, and registers also the changes. GIS basic skills are needed.

The test in the Bugala case shows that the use of handheld GPS systems to validate points proves to be more cost-efficient than having to rely on conventional surveying methods. In connection with the use of publicly available satellite and/or aerial images, both the organization spatially referencing points seem relatively cheap. The basic georeference also allows for scaling up and linking to other basic registers, which shows improvements in the accessibility of the contained information. In addition, the use of visual images makes it easier to connect to stakeholders and to the public than when using conventional systems or cadastral parcel plans.

Technically, the preliminary conceptual data model for the Bugala case can be further explored. In cases where local utilities and technical infrastructure is insufficient, the database itself could be stored in the cloud. This would prevent the risk of failure in case of power cuts or insufficient storage capacity in local areas. The storage of geospatial elements with point still has the potential for upscaling and linking to other databases and/or registers. The tests have shown the potential for scale-up the small design to either a multipurpose cadastres or a system contained and maintained in a cyber-information infrastructure. Even without a GPS in the field, using a paper aerial or satellite image and a pen, it would be possible to register points and later transfer these points in a digital environment on completion of the fieldwork. In this case, no specific knowledge of GPS systems would even be necessary to build up and populate a reliable database.

Despite the moderately positive results obtained through the two cases there are still some limitations when constructing a point cadastre. An obvious one is that the full utilization still needs time before any benefits become visible. Though the development of a point cadastre may be a good start to enhance the explicitness of tenure, tenants, and claims on land, one would need to evaluate who will profit most from using this information after several years. This will largely depend on how activities are connected to the new system and whether the system becomes embedded in daily institutional routines. The discontinuation of the actual construction and testing of the next prototype in Bissau given the instability of government also shows the dependence on the sociopolitical environment in which land management can be executed, and a cadastre needs to be built.

With the emergence of social media, the ability to upload spatially referenced information via mobile-/smartphones adds another potential technical element in the development of point cadastres. Such tools can indeed contribute to *quick-and-dirty* registers, even when the data quality or completeness of attributes needs to be improved over time. The main lesson of building point cadastres is that geometric quality is not of utmost importance if no information is available at all. Instead, reasoning from uptake and usage of information and adaptability to local context is more relevant in such cases. The acceptance of reduced geometric quality does have consequences for the land surveying profession in the phase of setting up the basic register. However, once the basic system is in place, the main task becomes updating and maintenance. Given that in a country such as the Netherlands there are 120,000 geometric updates per year (on a total of 10 million parcels), there is still a lot of work to do for para land surveyors or fully accredited/licensed land surveyors. However, in a country such as Guinea-Bissau, where there is historically a limited number of professional surveyors, one is required to rely on para professionals, who can within a relatively limited time be trained in the basic acquisition and recording of points and attributes.

Conclusion

The main issue of this chapter is whether point cadastres could be an alternative in case of poor or incomplete land administration systems. The two cases and related prototypes have shown how to build a point cadastre and have identified several challenges to overcome in this process. In each case, it has been possible to construct a basic database and connect this with a georeference using an online base map. Regarding the challenges, these include technical challenges of fit-for-purpose data models and data acquisition procedures, organizational challenges of retrieving the data with paraprofessional human resources, institutional and legal challenges of either adapting or aligning with current legislations, and social challenges of ensuring the involvement and acceptance of local stakeholders. A technical challenge also still remains: the access to technology, basic utilities, and infrastructure and technical skills. In addition, it is important that the design does not occur in isolation of the sociopolitical and physical context. Within these boundaries, the conclusion is that it is possible to link sufficient and appropriate attribute information about a parcel and tenure via a point and that such a method may contribute to responsible land administration.

The two cases have also revealed some difference in the design for urban versus rural point cadastres. The technical specifications are probably easier to reach in rural areas as compared to urban areas. Objects and areas in rural are often easier to identify and pinpoint by a single point in case of limited overlapping structures or in case of obvious boundaries. At the same time, however, if there are no overlapping trees, the case of Guinea-Bissau has shown that even in densely populated urban areas it is relatively easy to pin-point roofs. A crucial difference is perhaps more related to its use. In an urban setting, the main interest and usage is likely activities such as tax collection, whereas in rural areas it mainly supports tenure security. Finally, the physical conditions may play a role. It works well in a city such as Bissau, where there are no high-rise buildings and the area is relatively flat. This make the construction of a point cadastre feasible and appropriate, even though technically one could opt for more than one dot per roof, if there were multiple rights to pinpoint. For this specific case however, there existed a culture where boundaries of areas were hardly contested. Similar conditions would apply for informal settlement areas or cities with similar physical characteristics or similar institutional practices. This would be much more complicated in densely populated cities with many high-rise buildings or large boundary conflicts. In such cases, a point cadastre would not facilitate data collection nor tenure security.

Unlike Western/Northern paradigms in land administration spatial quality receives perhaps less priority in the design of the cadastral system, thus favoring quality factors such as fit-for-purpose, equality of tenants, and complementarity and accountability of the system holders. Such an approach is especially suitable where interests in land are unequal or opaque, where undocumented tenure does not have the same weight as formal registration systems, or where connections to other basic registers, such as the tax and civil registers, are lacking.

With regard to the methodology of designing and applying technology, the tests have proven that the choice for requirements engineering seems appropriate in the design and implementation of *responsible* point cadastres. It not only develops the technical system based on certain requirements, but it also forces the system designers to systematically connect with stakeholders to collectively formulate design requirements, test and validate such requirements, and test and validate first prototypes before actually relying on such systems. In locations where there is limited capacity in surveying legal matters and ICT, the gradual design solution can thus be a responsible solution.

As the prototypes could not further be implemented it is recommended that the point cadastre approach could be further tested in other countries and contexts. This could include looking into linkage to other types of registers as well.

References

Burke, L. 1995. *Urban and Municipal GIS Applications in Developing Countries—The Problems and the Potential*. Paper presented at the ESRI user conference, Wyndham Hotel and Convention Center, Palm Springs, CA, May 22–26.

Davies, C., and C. Fourie. 2002. A land management approach for informal settlements in South Africa. In *Holding Their Ground: Secure Tenure for the Urban Poor in Developing Countries*, edited by A. Durand-Lasserve and L. Royston, pp. 218–230. London, United Kingdom: Earthscan Publications.

de Vries, W.T., J. Lewis, and Y. Georgiadou. 2003. The cost of land registration: A case study of cost efficiency in Namibia. *The Australian Surveyor* 48(1): 7–20.

Deininger, K., C. Augustinus, S. Enemark, and P. Munro-Faure. 2010. *Innovations in Land Rights Recognition, Administration, and Governance*. Washington, DC: World Bank Publications.

Enemark, S., and I. Williamson. 2004. Capacity building in land administration—A conceptual approach. *Survey Review* 37(294): 639–650.

Fourie, C. 1994. *Options for the Cadastre in the New South Africa: Report to the South African Council for Professional and Technical Surveyors*. KwaZulu-Natal, South Africa: Department Surveying and Mapping, University of Natal.

Hackman-Antwi, R. 2012. *Design and Assessment of a Procedure for Building and Maintaining Point Cadastres*. Enschede, The Netherlands: University of Twente, Faculty of Geo-Information and Earth Observation (ITC).

Hackman-Antwi, R., R.M. Bennett, W.T. de Vries, C.H.J. Lemmen, and C. Meijer. 2013. The point cadastre requirement revisited. *Survey Review* 45(331): 239–247. doi: 10.1179/1752270612y.0000000015.

Hanstad, T. 1998. Designing land registration systems for developing countries. *American University International Law Review* 13: 647.

Home, R., and J. Jackson. 1997. *Our Common Estate: Land Rights for Informal Settlements: Community Control and the Single Point Cadastre in South Africa*. London, United Kingdom: The Royal Institution of Chartered Surveyors.

Keuber, S. 2014. *Validation of Point Cadastre Requirements in Practice: The Case of Bugala Island, Uganda*. Enschede, The Netherlands: University of Twente, Faculty of Geo-Information and Earth Observation (ITC).

Tuladhar, A.M., M.J.M. Bogaerts, and P. van der Molen. 2004. *Parcel-Based Geo-Information System: Concepts and Guidelines*. PhD thesis, ITC Dissertation 115. Enschede, The Netherlands: ITC.

van Vliet, H. 2008. *Software Engineering: Principles and Practice*. 3rd ed. Chichester, West Sussex, England: John Wiley & Sons Ltd.

8

Digital Pen Method

Didier Milindi Rugema, Jeroen Verplanke, and Christiaan Lemmen

CONTENTS

Introduction

One innovation in cadastral data collection is in using satellite images and drawing boundaries in the field with land right holders as witness (Lemmen and Zevenbergen 2010). This chapter presents an evaluation of the use of a digital pen method as an example of a new unconventional approach in cadastral data acquisition. Conventional approaches, often of historical footing, are inadequate in many jurisdictions. For example, highly rigorous and accurate methodologies and procedures, practiced by registered or licensed surveyors, are characterized by long duration and delays in completing acquisition. These delays are represented by insufficient coverage of the registered land, and the required accuracy leaves much potential for errors. As these are not pro-poor approaches, alternatives to this mighty accuracy tradition in land administration are needed. Flexibility is needed in relation to the way of recordation, the type of spatial units used, the inclusion of customary and informal rights, the data acquisition methodologies, and in the accuracy of boundary delineation. It is less important to produce accurate maps. It is deemed more important to have a complete cadastral map and to know how accurate that map is (Lemmen and Zevenbergen 2010; Lemmen 2012). The hypothesis in this chapter is that the use of a digital pen method allows cadastral data acquisition can to be done in less time, with the same number of people, whilst preventing duplication of errors.

First, the participatory mapping approach for cadastral boundary mapping with the use of satellite imagery is presented. The technology behind the digital pen is described before an overview of the use of the digital pen is given on the basis of a study done in Rwanda (Rugema 2011). The results of this study are finally compared to an existing method for field data collection in Rwanda. These tests with the digital pen and the described outcomes add to the practice-based knowledge of participatory mapping and show under which circumstances the technology can be an alternative for conventional forms of cadastral surveying and data acquisition.

Theoretical Perspective

Participatory Mapping

One of the approaches to register land is the involvement of local people in the process of land registration through participatory mapping. The community involvement in land tenure regularization activities is seen as an instrument for engendering social capital and a strategy for resource mobilization toward securing tenure (Magigi and Majani 2006). In South Africa, in the Mpumalanga Province local people traced their views on how land reform would be done in their area on paper maps by using pencils and color markers (Weiner and Harris 1999). The value of participation in land registration is the agreement by local people on the consolidated information. For this reason, the first phase is to share public information. In Cambodia, land is registered systematically by using ortho-photos and printed maps. Local people participate to provide information about their parcels, and the adjudication officer and the demarcation officer literally go together to the parcel in question on an agreed date. During land adjudication, the existing rights to parcels are ascertained, not altering the existing rights or creating new ones. In demarcation of parcels, boundaries are delineated and agreed on with the adjoining owners or other interested parties (Törhönen 2001).

Transparency is important in participatory mapping to protect the rights of all parties. According to Weiner, Harris, and Craig (2002), "Community-based GIS projects simultaneously promote the empowerment and marginalization of socially differentiated communities. As a result, the nature of the participatory process itself is critical for understanding who benefits from access to GIS and why" (p. 3). According to Törhönen (2001), special attention has to be paid to vulnerable groups such as women or poor, disabled, or illiterate people. If this is not recognized, powerful people could take advantage and formalize land grabbing. Prevention comes in this case through transparency in the form of publicity.

Cadastral Data Acquisition

A cadastral map represents boundaries of ownership or land use rights, for example, customary land rights, or informal land rights as possession or occupation (Lemmen 2012). Those boundaries should represent the actual situation as it exists in the field. The cadastral map is, in fact, a map visualizing that people agree on the boundaries of their properties, living areas, or living environment. Even disagreement (overlapping claims) can be mapped. In this respect, it can be seen as a social map. It can also be seen as a map representing legal certainty in relation to ownership or factual land use. The map can be used as a basis for land tax; this is again a social issue in relation to the contribution of individuals, families, or groups to the development of society.

McCall (2003) argues that untrained people, with local spatial knowledge, can work effectively, easily, and happily interpreting aerial photos. Ortho-photos and satellite imagery may very well be used as a basis for data acquisition in the field of cadastral boundary data collection (Lemmen and Zevenbergen 2010). Cadastral boundaries can be identified in the field on top of such images. Identifiers of spatial units or parcels have to be included.

Using satellite imagery for cadastral applications is not new (Konstantinos 2003; Tuladhar 2005; Kansu and Sezgin 2006; Ondulo and Kalande 2006). Satellite images have been used for cadastral boundary acquisition in earlier doctoral research in Ethiopia (Haile 2005), as well as in pilot projects in Rwanda (Sagashya and English 2009) and Namibia (Kapitango and Meijs 2009). In Ethiopia, in 2008 a team conducted a simple field test using QuickBird satellite imagery in the program of rural land certification. The results showed that high-resolution imagery based on land adjudication is useful in participatory mapping. The data collection in the field was done with the help of land rights holders and local officials. The image quality of the plots at a scale of 1:2000 was sufficiently high to allow the parties to easily understand the images and contribute input, making the process very participatory (Lemmen and Zevenbergen 2010). Since 2013, Ethiopia has been registering land parcels systematically at a national level based on the experiences of Rwanda.

Cadastral maps can be used as a basis for spatial planning. Before changes are implemented, the existing situation on formal, informal, and customary land rights needs to be known. In combination with a registry of landowners or users, the map provides a basis for access to credit, for example, mortgage or micro-credit; further, the identified land rights can serve as collateral.

Often, a distinction is made between "general" and "fixed" boundaries; see the studies by Henssen (1995) and also Bogaerts and Zevenbergen (2001).

Henssen states that the English system mainly relies on physical boundary features: man-made or natural. The precise position of the boundary within these physical features depends on the general land law of the country concerned. This system is called the "general boundary system." Inclusion of the survey data in the cadastre implies the boundary to be "legally fixed." In some land administration systems, the location of the boundaries is guaranteed. According to Henssen, the choice between fixed and general boundaries depends on the pace of creating or updating the system, the existence of physical features, disputes to be expected, the amount of necessary security, and costs. Important observations in the field may be to identify to whom the physical boundary belongs. Fixed boundaries are based on observations (field surveys) in the field. Cadastral boundary measurements can result in calculated or observed (and transformed) coordinates, which are inputs for a cadastral mapping process. Enemark et al. (2014) in their joint FIG/World Bank publication on "fit-for-purpose" land administration highlight that land administration systems—and especially the underlying spatial framework of large-scale mapping—should be designed for the purpose of managing current land issues within a specific country or region, rather than simply following more advanced technical standards. The fit-for-purpose approach is flexible, participatory, inclusive, affordable, reliable, upgradable, and hence ultimately responsible.

Features of the Digital Pen

In the existing analogue method of general boundary survey, ortho-photos are first plotted and then boundaries of parcels are drawn on site and the parcel identifiers (IDs) are written with a pencil on field sheets. Post-processing has to be performed in the office. This concerns redrawing and rewriting over the pencil marks by using a normal pen on field sheets, scanning and geo-referencing field sheets, and vectorising parcel boundaries. This process induces different sources of errors because of many process steps. It is double work in terms of time. It requires a lot of space for archiving. This implies complex information management.

The innovation of technology has changed the ordinary pen into a smart pen called a digital pen that can directly record annotation in both analogue (ink-paper-based) and digital (computer-based) formats. The first modern digital pen was already released in 1996 by Anoto, a company based in Sweden. Anoto developed an ink pen equipped with a digital camera that takes snapshots to transfer ink into digital data. The digital pen works in conjunction with a special "digital paper," which is imprinted with dot patterns called Anoto patterns. The digital pen with digital paper by Anoto has become the most widely used standard for the digital pen technology (Schneider 2008).

On the basis of the Anoto standard, third parties have further developed the digital pen technology, which can be used in many applications including geospatial data collection. For this purpose, the technology was added as an extension to ArcGIS (ESRI Inc.), but dedicated solutions have also been built to function with other geographic information system (GIS) software. In this chapter, however, we refer to the solution provided by Adapx, which consists of a software extension for ArcGIS version 9. The digital pen for geospatial data collection has a capability to directly record spatial data (points, polygons, and lines) with predefined attribute data directly in digital format. With this capability, data can be processed with both computer-aided design and GIS software as a geo-database can be populated almost automatically after field data collection.

The pen's principal work is similar to that of a scanner. Each stroke of the pen in the paper consists of writing, scanning, and digitizing instantly. The digital sensor (camera) automatically scans the movement of the pen in conjunction with the pattern on the paper at a rate of 75 frames per second (Roe 2009). The pattern consists of numerous tiny black dots (0.1 mm in diameter), visible as a gray haze as they are arranged with a spacing of approximately 0.3 mm (Livescribe 2010). Each paper has a unique pattern with its unique combination of dots in every small area. The digital pen records the feature written in its exact position based on the dot's position in the paper. The pattern is printed in a black ink that reflects infrared light (800–950 nm) to be recognized by the pen's sensor. The accuracy when using the digital pen ranges up to around 0.1 m which is the effect of two conditions:

1. The Anoto pattern resolution is 0.3 mm.
2. The maximum error calculated by the location of the pen and the dot pattern is 0.7 mm (variation dependent on the angle of the digital pen against the paper).

The pattern can be printed on different types of paper that meet Anoto's pattern-enabling requirements such as opacity, reflection, surface roughness, and weight. It is possible to print satellite images or photos with a pattern. The printed image can then be brought to the field to draw the boundaries of spatial units with different land (use) rights. The superimposed Anoto pattern, however, reduces the contrast and definition of the printed image. The drawn boundaries can be loaded to a computer or transferred via Bluetooth and can be automatically overlaid on the maps or digital images available in GIS software or Google Earth.

Methodology

Rugema (2011) tested the digital pen in Rwanda. Three main aspects were investigated with respect to the use of a digital pen method. First was the acquisition of polygons, lines, and points of boundaries with the digital pen in the real-life situation of Rwanda. Second, the evaluation of these tests was done on the basis of the conventional approach used in Rwanda to understand how its use would affect the time required for data acquisition and what sacrifices would be made with respect to accuracy. Finally, it was assessed whether the method could be an appropriate alternative to the existing analogue method to achieve national coverage in accordance with current government policy.

Land registration in Rwanda has been done in a systematic way. In May 2013, about 10.4 million parcels were registered and 8.8 million printed land lease certificates were issued (see the case study on Rwanda by Enemark et al. [2014]). In the existing ortho-photo-based land adjudication, a pencil is used for data collection in the field on plotted maps for demarcating boundaries and writing parcel IDs. After fieldwork, the data are post-processed by first redrawing and rewriting the pencil marks on the field sheets with a normal pen to make the boundaries and parcel IDs more visible for scanning. Second, the field sheets are scanned and geo-referenced, and the drawn boundaries are digitized. This process implicates different sources of errors because of the many steps involved in it. It is double work (redrawing and digitizing), and a lot of space is required for archiving the field documents. The method, however, facilitates participatory mapping as stakeholders can "sit around the map" while constructing it. Data collectors do not need much training or education; anyone who is able to interpret the image, and can read and write, can do the job. The tools are easy to use, are suitable for field conditions, and facilitate collaboration in the field: the printed field sheets are on a large or small scale, and they are lightweight, portable, and reliable for fieldwork. Despite these advantages, there is a need to investigate alternative data acquisition methods to prevent the sources of errors that are present in this existing method of analogue field data collection and to improve efficiency (Rugema 2011).

The digital pen was tested in the field to assess whether it could reduce errors and improve efficiency. It captures what is drawn and written on digital paper, and it stores the drawn and written data to its internal memory and the operator can transfer the collected data to the computer by connecting the pen via a Universal Serial Bus (USB) adapter. The digital paper in the test included a legend: when touching a feature in the legend with the pen, it stores this as a classification for the next drawn information on the digital paper. For annotations on the map, a markup layer can be selected in the legend, and the operator can thus write as much extra information as needed on the field map next to the boundaries. The annotation on the map is used for instance to mark a parcel with the appropriate ID, indicate mistakes, and give remarks. Annotations are visualized as a separate layer in GIS after uploading the data. Within the used software (Capturx for ArcGIS), these annotations are unfortunately stored as vector information and not as digital attributes that can be displayed in a spreadsheet. This means that the parcel IDs must still be entered manually into the attribute table of the geo-database.

The following materials were used during field testing:

- ArcGIS and Capturx for ArcGIS software for producing and printing maps as digital paper.
- Penx digital pen (by Adapx) with USB adapter.
- QuickBird satellite imagery (0.60 m resolution) for the Nyamugali Cell, in Gatsata Sector/Kigali City.
- Ortho-rectified aerial photos/ortho-photos (0.25 m resolution) as used in the existing method for the land adjudication process at the national level.
- Plotted ortho-photos printed on A3 paper size at scales 1:1000 and 1:1500, respectively, for the test areas in Musezero and Kibonga. These scales were according to the mapping scales of the existing method in these two areas.

Results

Polygon-Based Feature Mapping

If boundaries are drawn in polygon mode, a parcel should be drawn in a continuous stroke of the pen from start to finish of the boundary. Each lift of the pen implicates the start of a new polygon. If a polygon drawn on digital paper is not closed (the end point being not near enough to the starting point), then the polygon will be closed by the application software in an unpredictable way. Therefore, when drawing parcel boundaries in this mode the operator has to pay close attention to the "closing" of the polygon.

Figure 8.1 presents two cases of drawing boundaries of neighboring parcels. On the left side of the figure, two polygons are drawn as completely closed polygons. The parcels are mapped individually: the result is barely visible gaps and overlaps in the shared boundary. In the right-hand image, the polygon drawn on the right has been inserted by connecting it to the polygon on the left. The result in this case is that the connected (not-closed) polygon is closed in an incorrect way by the software. The used software (Capturx 1.2 for ArcGIS) did not have the right algorithms included to properly auto-complete the parcels adjoining the polygon. To share the boundary between the two parcels, post-processing in ArcGIS was required.

The digital pen uses ink (as a ballpoint pen), and everything drawn and written on the map is therefore permanent; it cannot be erased from the field sheet as it would be when using an ordinary pencil. The available option for the digital pen to make corrections is by making additional annotations on the map to indicate any mistakes.

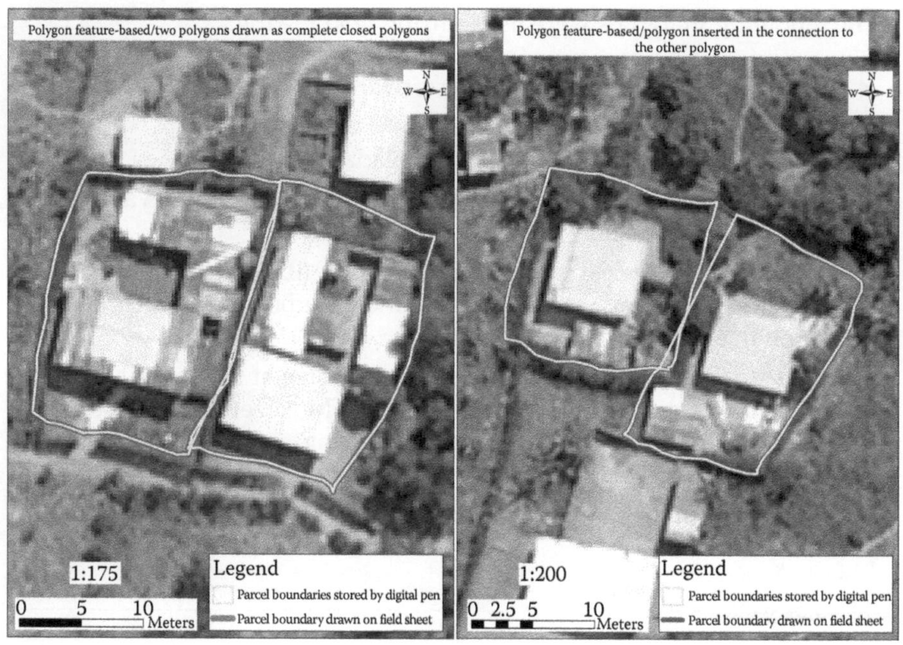

FIGURE 8.1
Polygon feature–based representations of boundaries between neighboring parcels. (From Rugema, D.M., *Evaluation of Digital Pen in Data Capturing for Land Administration Purposes in Rwanda*, Faculty of Geo-Information and Earth Observation-ITC, University of Twente, Enschede, The Netherlands, 2011.)

Polygon-based feature mapping requires objects to be collected as complete polygons. As the object of observation in the field is a boundary, a line feature, it was deemed a better option to build parcel boundaries out of individual line segments. As polygons would be created anyway through (automated) post-processing when all boundaries in the perimeter are drawn, there is no need to draw these polygons in the field.

Line-Based Feature Mapping

Demarcation in the field implies observation of the boundary in the field to draw it on a map as a representation of how it appears on the ground, and how it connects to other boundaries. When a boundary perimeter has been mapped, a para-surveyor issues a unique consecutive number to a parcel. With the digital pen, using a line-based mapping mode, it is possible to draw a single line segment, move the pen from the map, assess its quality, check the following boundary on the ground, and continue this exercise until the completion of the perimeter.

When mapping line segments, the operator must make sure the segments start close enough (tolerance set by software) to connecting/existing lines.

If the lines are overlapping, these overlaps can create small new polygons that do not represent any parcel existing on the ground. Figure 8.2 shows those kinds of errors.

The errors shown in Figure 8.2 (right) are usually not visible on paper. They become visible after uploading data from the digital pen into a GIS and zooming in to a submeter scale. Figure 8.2 shows the connection of four parcels existing on the ground (left image). When the line features were converted to polygon features, this only applied to one parcel. The other three parcels could not be formed because of the gaps between the connecting lines. In addition, however, four new "ghost polygons" were created connected to those four parcels. These extra polygons are very small; they are only visible on a screen at a higher scale than the one used in data collection. In reality, these nonexistent polygons must be tracked and edited out in post-processing. These kinds of extra polygons happen by mistake when the digital pen touches the digital paper unintentionally or is slightly moved on starting or finishing a line segment. The main reason for these problems is the streaming digitization mode of the digital pen. When drawing in streaming mode without the option of point mode digitization, the precision by which the operator can draw parcel boundaries on digital paper becomes

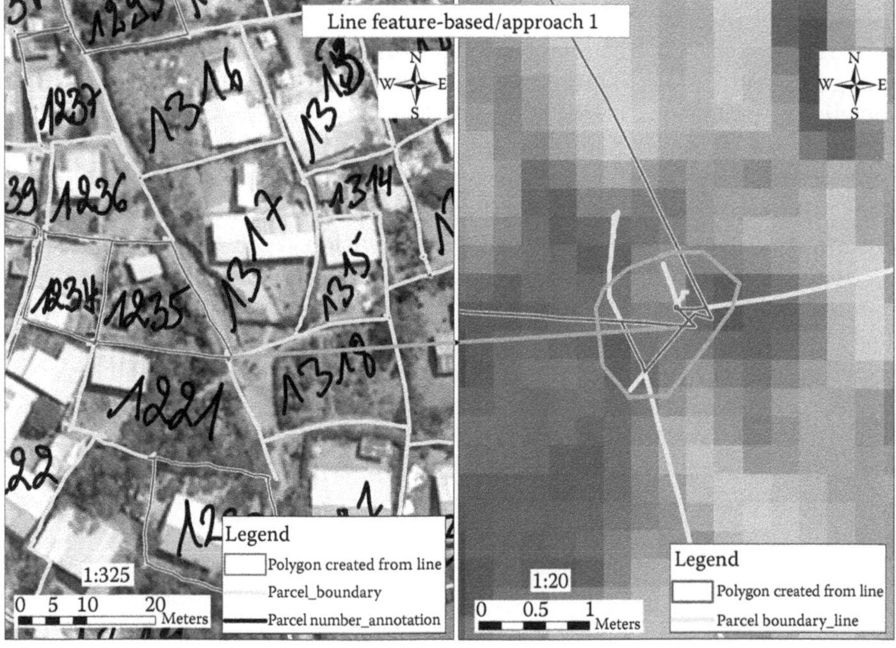

FIGURE 8.2
Errors in a line-based feature approach. (From Rugema, D.M., *Evaluation of Digital Pen in Data Capturing for Land Administration Purposes in Rwanda,* Faculty of Geo-Information and Earth Observation-ITC, University of Twente, Enschede, The Netherlands, 2011.)

less; thus, the control over the mapped coordinates becomes less. This causes a redundancy of vertices (vector coordinates) for parcel boundaries at every vertex; each turn of the pen on paper (sometimes representing field distances ranging up to around 0.1 m) is recorded as part of the boundary. Manual checking of errors and making corrections are time consuming and difficult. Fortunately, much of the errors and redundancy can be automatically removed in post-processing by setting tolerances for line segments. According to the errors occurring with this approach, and the time it takes to correct these errors, a point-based approach was tested as well to assess the way in which the digital pen can capture cadastral data.

Point-Based Feature Mapping

The quality of line-based mapping depends much on the steady hand of the operator, particularly when drawing relatively long lines on paper. With point mode digitization, parcel boundaries can be drawn with control over where the digital pen makes its mark on the map, thus with better control of resulting coordinates (vertices). During this test, point features were marked at the corner points (change of vector angle) of a parcel boundary. The resulting point map would, however, be difficult to interpret in the field. Therefore, each parcel boundary line would also be drawn on the map as markup (annotation).

In ArcGIS, the digitization of parcel boundaries was done by creating a polygon feature class and snapping to the collected points uploaded from the digital pen. Annotated parcel IDs (Figure 8.2, left) were then used to complete the attribute data that uniquely identify each parcel. The output then required manual post-processing to close any incomplete parcel polygons. This requires careful "snapping" of the points to give two neighboring parcels the same vertices for their line of division. With the use of tailored software for the digital pen, the order of point processing to form unique parcels can be automated and executed to avoid this post-processing step. The digital pen stores sequential steps in mapping through which corresponding IDs can be provided to point features belonging to the same parcel. The additional requirements to enable this are a more elaborate field form to be printed on the digital paper map and a consistent mapping approach by the operator, as only points belonging to one parcel can be mapped at a time and shared points must be indicated separately.

Discussion

The digital pen method as tested in Rwanda is almost similar in process to the actual participatory approach used in Rwanda (Rugema 2011). Local people could therefore easily familiarize themselves with this new method. The essential part is the ability to interpret images for parcel boundary demarcation. With the digital pen method, the main difference is that the boundaries are additionally collected as geo-referenced digital features. The point feature-based mapping approach appears to be the most practical. It creates the least number of errors, and it reduces the amount of post-processing. As a consequence, the digital pen could reduce the time required for cadastral boundary survey. However, the point feature-based approach does require additional tailored software for automated post-processing of parcels' boundaries. Compared to the existing method, the point-based method also gives a similar or more accurate output and in less time. Acquisition of cadastral data directly into a geo-referenced digital format can, based on these tests, be useful to reduce the number of steps required to get the final digital land information. This has an impact on reducing errors and saving time as it will reduce the many steps involved to achieve the final digital outputs in a conventional way.

However, the digital pen has been used in practice for cadastral boundary data acquisition purposes in "low technology, low internet environments" where it eventually mounted into frustration (Glenn K., pers. comm.; Thomson N., pers. comm.). Next to solvable technological issues as experienced in Rwanda, there were reported download problems and compatibility issues with GIS software that were not conducive to easy use and maintenance (Thomson N., pers. comm.). In addition, the main problem with the digital pen seemed to be the time needed to "clean" the data after they were uploaded, particularly the unwanted points, lines, and polygons that did not always close properly (Glenn K., pers. comm.). Although most of these errors are user errors that can be reduced by training, when working with local staff, it underscores the difficulty of dealing with the pen's sensitivity. The issue is that the technology needs to be "plug-and-play" in all aspects (preparation, mapping, and post-processing), which is not the way the technology "comes out of the box." The users reported, "We were losing too much sleep over trying to get it to work so we dropped it" (Thomson N., pers. comm.).

Luckily, many of the experienced problems can be solved by using more tailored software solutions with a dedicated workflow attached to the digital paper maps. Our cases indicate the feasibility of the tool for general boundary mapping in rural areas and as an updating tool for more general land administration issues. Further research is needed to develop the required innovative software to enlarge the domain of application of the tools. We suggest that with little adaptation it can be useful for small-scale mapping (e.g., large farming/plantation areas) as well as in urban areas for both initial data collection and updating of existing databases. More comparative studies are recommended for the digital pen with other data collection methods and to test the digital pen to other evaluation criteria, particularly more user-driven criteria.

Evaluation of the pen should focus on possible improvements based on those field experiences. The digital pen method would be more practical by improving some functionality of the techniques related to this method, particularly in the use of line feature based and/or polygon feature based in the field. The software available for the tools needs to be improved, and further testing has to be done. The options to be added are snapping option and point mode digitization. Another option, for any feature type adopted, would be the possibility of erasing the errors done on a digital printed map and this to be related to the information/data stored digitally into the memory of the digital pen. This would improve the quality of final outputs and the saving of time for office workload. However, a successful large-scale application of this tool in general boundary mapping is required before this innovation can be expected to fill the gap of low-cost high-tech data acquisition in land administration.

Conclusion

In conclusion, we can argue that, even though the technology is not yet perfected, the digital pen method has potential for (general) parcel boundary registration in rural and urban areas. It has the ability to collect large quantities of parcel-related data in a participatory manner while reducing the time and workload of post-processing these data. It can therefore assist the establishment of responsible land administration systems. However, data collection using the digital pen should be optimized according to its functions and purposes. Some recommendations to optimize the pen are using point features to get better geometric accuracy and reducing the post-processing time by adding algorithms and dedicated workflow for post-processing into the software.

The most important contributions of the digital pen method for updating cadastral maps are its functionality and usability to provide geo-referenced digital raster (originally drawn) data. The digital pen can take a role for demarcating preliminary boundaries before (if necessary) conducting higher accuracy measurements. The advent of digital pen technology therefore coincides with the urgent need for new approaches in land administration and land management where conventional approaches prove to be inadequate in rural and traditional environments or where flexibility is needed in relation to the way of data acquisition methodologies and the accuracy of boundary delineation. Rather than having an incomplete and highly accurate cadastral map, it is more important to have a complete cadastral map and to know how accurate that map is.

References

Bogaerts, T. and J. A. Zevenbergen. 2001. Cadastral systems—alternatives. *Computers, Environment and Urban Systems* 25(4–5): 325–337.

Enemark, S., K. C. Bell, C. H. J. Lemmen, and R. McLaren. 2014. *Fit-For-Purpose Land Administration*. FIG publication no. 60. Copenhagen, Denmark: FIG/World Bank Publication. ISBN: 978-87-92853-11-0.

Haile, S. A. 2005. *Bridging the Land Rights Demarcation Gap in Ethiopia: Usefulness of High Resolution Satellite Image (HRSI) Data*. PhD diss. Vienna, Austria: BOKU University of Natural Resources and Life Sciences.

Kansu, O. and G. Sezgin. 2006. *The Availability of the Satellite Image Data in Digital Cadastral Map Production*. XXIII International FIG Congress. Munich, Germany, October 8–13.

Kapitango, D and M. Meijs. 2009. *Land Registration Using Aerial Photography in Namibia: Costs and Lessons*. FIG-WB Conference: Land Governance in Support of a Domain Model for Land Administration. Washington, DC: World Bank.

Konstantinos, C. 2003. *Combination of Satellite Image Pan IKONOS-2 with GPS in Cadastral Applications*. UNECE WPLA Workshop on Spatial Information Management for Sustainable Real Estate Market: Best Practice Guidelines on Nationwide Land Administration. Athens, Greece.

Lemmen, C. H. J. 2012. *A Domain Model for Land Administration*. PhD thesis. Delft, The Netherlands: Delft University of Technology. ISBN/EAN 978-90-77029-31-2.

Lemmen, C. H. J. and J. A. Zevenbergen. 2010. First experiences with high-resolution imagery-based adjudication approach in Ethiopia. In: *Innovations in Land Rights Recognition, Administration and Governance*: edited by K. Deininger, C. Augustinus, S. Enemark, and P. Munro-Faure, 96–104. Washington, DC: World Bank.

Livescribe. 2010. Introduction to the Livescribe Platform. Accessed October 2014. http://www.livescribe.com/en-us/media/pdf/dev/Livescribe_Platform_Introduction.pdf.

Magigi, W. and B. B. K. Majani. 2006. Community involvement in land regularization for informal settlements in Tanzania: A strategy for enhancing security of tenure in residential neighbourhoods. *Habitat International* 30(4): 1066–1081.

McCall, M. K. 2003. Seeking good governance in participatory-GIS: A review of processes and governance dimensions in applying GIS to participatory spatial planning. *Habitat International* 27(4): 549–573.

Ondulo, J-D and W. Kalande. 2006. *High Spatial Resolution Satellite Imagery for PID Improvement in Kenya*. XXIII FIG Congress: Shaping the Change. Munich, Germany.

Roe, D. 2009. Adapx Captures and Converts Handwritten Data into SharePoint 2010. Accessed October 2014. http://www.cmswire.com/cms/document-management/adapx-captures-and-converts-handwritten-data-into-sharepoint-2010-005830.php.

Rugema, D. M. 2011. *Evaluation of Digital Pen in Data Capturing for Land Administration Purposes in Rwanda*. Enschede, The Netherlands: Faculty of Geo-Information and Earth Observation-ITC, University of Twente.

Sagashya, D. and C. English. 2009. *Designing and Establishing a Land Administration System for Rwanda*. Technical and Economic Analysis. FIG–World Bank Conference: Land Governance in Support of the Millennium Development Goals: Responding to New Challenges. Washington, DC: World Bank.

Schneider, K. 2008. The Re-birth of Pen and Paper in Mobile GIS. Accessed October 2014. http://gislounge.com/the-re-birth-of-pen-and-paper-in-mobile-gis.

Törhönen, M-P. 2001. Developing land administration in Cambodia. *Computers, Environment and Urban Systems* 25(4–5): 407–428.

Tuladhar, A. M. 2005. *Innovative Use of Remote Sensing Images for Pro-Poor Land Management*. FIG Expert Group Meeting on Secure Land Tenure: New Legal Frameworks and Tools. Bangkok, Thailand.

Weiner, D. and T. M. Harris. 1999. *Community-Integrated GIS for Land Reform in South Africa (No. 9907)*. Morgantown, WV: West Virginia University, Regional Research Institute.

Weiner, D., T. M. Harris, and W. J. Craig. 2002. Community participation and geographic information systems. In: *Community Participation and Geographic Information Systems*, edited by W. J. Craig, T. M. Harris, and D. Weiner, 3–16. London, United Kingdom: Taylor & Francis.

9

Accelerated Land Administration Updates

Mireille Biraro, Rohan M. Bennett, and Christiaan Lemmen

CONTENTS

Introduction

Land administration is about determining, recording, and disseminating land information (UNECE 1996). Land information is a fundamental ingredient in state-backed land tenure systems, valuation systems, land use planning, and land development (Enemark 2004). In most developing countries, land information is created as part of large countrywide land registration projects financially supported by international donors. However, creation must be followed by maintenance: lasting effects on tenure and land market facilitation require that systems keep running long after donors leave (Magis and Zevenbergen 2014). To remain useful, a land administration system must reflect the reality on the ground and this is only possible when all changes in land information are reported (Zevenbergen 2002). Indeed, Williamson et al. (2010) assert that if the changes are not captured in state-backed land registers, the system loses societal relevance and is eventually replaced by an informal system.

In the early 2000s, Rwanda began the process for establishing a land administration system: all rightful claimants would be provided legally valid land documents through a systematic land registration process. The land information was collected in a participatory manner: parcels were surveyed by grassroots surveyors using aerial images and a general boundary approach. The resultant land information was gathered into a digital land register: the Land Administration Information System (LAIS). The LAIS was intended to support the maintenance phase. While the initial land registration was systematic, the updating process was sporadic: transacting parties would come on an individual basis to district land offices and report changes. Requirements were defined and workflows designed for the various forms of land transaction.

Biraro (2014) analyzed Rwanda's updating procedures and identified obstacles that could discourage the reporting of changes. These included (1) long updating processes due to the analogue nature of the system, the lack of technology used in service delivery, the surveying approach that combined fixed and general boundaries, and the many processes that had to be followed for a single transaction (e.g., to donate a part of land parcel, the right holder follows two processes) and (2) the high registration fees of rights transfer compared to the value of land dealt with in the transaction. It was hypothesized that both obstacles could be better understood and potentially overcome by using the concepts and tools of workflow management, specifically the use of the Unified Modeling Language (UML). Workflow management can be used to analyze and redesign the activities, actors, technologies, and interactions inherent in a complex system. The approach is regularly used in the business sector, particularly in developed countries; however, the application to state-based land administration systems in less developed countries is a more recent phenomenon. In the Rwandan case, the methodology could be used to analyze the activity flows, actors, and technologies in the land information–updating process.

Many authors agree on the importance of updating land information subsequent to initial registration programs; however, in both theory and practice less attention is afforded to ensuring that updating procedures are simple and cost-effective for both state and citizens. In other words, the characteristics of fit-for-purpose land administration, as espoused by Enemark et al. (2014), should be better incorporated into land information maintenance. To this end, this chapter focuses on the updating challenge found in Rwanda. It aims to use workflow management concepts and tools to understand system bottlenecks and redesign options, ones that will improve the abilities of both state and citizens to keep land information updated. A theoretical grounding on land information updating and workflow management is first provided. Subsequently, the methodology, based on a design research philosophy, is described. Results are presented primarily using UML visualizations and accompanying descriptions. Discussions on the outcomes and future research directions follow.

Theoretical Perspective

Regarding definitions, the term land register is used synonymously with the term land administration system, land registration system, or land recordation system. Land information, sometimes referred to as land records, is broadly considered to include land tenure, value, and use information and is considered a defining component of a land registration system. Updating is considered synonymous with data maintenance and refers to the process of changing textual and spatial elements inherent in land information. The process is considered to include activities, actors, and technologies. Workflow redesign is related to workflow management and is considered synonymous with system upgrading. Upgrading is different from updating (Scheu et al. 2000): upgrading deals with system redesign, whereas updating is about updating the land information in an existing system.

The importance of keeping land registers updated is well recognized. A land register loses value if the information it stores is not updated (Henssen 2010). The failure or success of a registration system is dependent on the completeness and promptness of reporting land information changes (Binns and Peter 1995). Without reporting, the system loses relevance and is replaced by an informal system (Williamson et al. 2010). Although these arguments are theoretically straightforward, keeping systems updated in practice appears more difficult. This is particularly the case in the context of large donor-support initial registration programs: ensuring system maintenance is often a secondary concern (Jing et al. 2013).

The updating challenge is broken down into different elements: reference is made to the distinction between the time that the updating process takes and the cost to register the change (Larsson 1991; FIG 1995; Chimhamhiwa et al. 2009; Enemark et al. 2014). The related issue of accessibility to land offices, land services, and land information provides another lens (Larsson 1991; FIG 1995; Henssen 2010; Deininger et al. 2012; FAO 2012). The quality of land information that leads to security of transactions is also covered by FIG (1995) and Chimhamhiwa et al. (2009). Other aspects including concentration and decentralization of land services may be added to the list: both influence the success of the updating process. These aspects help to better understand updating and can guide the assessment of the existing situation and requirements definition for designing improved workflows.

Workflow is described as a set of tasks organized to perform a business process (Zur Muehlen and Indulska 2010). The concept has been effective in specifying, executing, and monitoring the flow of tasks (Heloisa and Mitchell 1996). Workflow management systems enable automation, via technology, of the processes between people and the tasks of a business (Aversano et al. 2002). However, automation can reduce human contact and even lead to a lack of motivation: people might feel controlled (Aguilar-Savén 2004). Workflow design starts with the identification of system requirements, often on increasing user satisfaction (Todorovski and Lemmen 2007). Norman (1996) suggests that requirements should explain (1) what the system is supposed to do, (2) what the system should have to do functionally to perform well, and (3) what the users want the system to do for them. In cases where systems interact with citizens, efforts should be made to ensure involvement and cooperation. In the case of land information updating, obstacles impeding the process should be removed to ensure that changes are reported: this workflow has to be customer oriented. Meanwhile, subsequent workflow design and management processes include process modeling, process reengineering, and workflow implementation and automation (Mentzas et al. 2001). Process modeling requires identification of workflow elements useful to capture an abstract process into workflow (Mentzas et al. 2001).

Workflow is one of the nine most frequently used modeling techniques used in a business process (Aguilar-Savén 2004). Various modeling languages are used to represent process activities that need to take place, step by step, to perform an action (Zur Muehlen and Indulska 2010). UML is one of these languages; however, it was specifically developed for information systems (Glassey 2008). It comprises nine diagram types. One type, activity diagrams, describes the flows circulating between activities within a process (Glassey 2008). In land administration practice, the UML activity diagram has been used to model processes. Activity diagrams receive less attention in land administration literature; however, UML class diagrams are heavily used in research dealing with land administration data modeling (Lemmen 2012). At any rate, Chimhamhiwa et al. (2009, 2011) use workflow management to measure land administration business processes whose activities are dispatched into different organizations. Zevenbergen et al. (2007) use UML activity diagrams to model real property transactions across European countries with an aim of providing a comprehensive and comparable description. The use of UML to understand and redesign land information–updating workflows in less developed countries receives minimal attention in the published literature.

Methodology

The research was design in nature: in alignment with workflow management theory, the research activities included requirement gathering, workflow modeling (including "as is" and "to be" situations), and validation processes.

Requirement gathering was primarily undertaken during fieldwork in October 2013 in Musanze, one of the five districts in Northern Province, Rwanda. The district has 476,655 registered land parcels, among which 421,555 (88.4%) had complete information in the land register (summary report on land tenure regularization program of May 31, 2013). In Rwanda, there are 24 recognized land administration workflows. These were developed as part of the Land Tenure Regularization program, a nationwide program that mapped and registered all landholdings, as recognized in new national land laws, over a 4-year period. For this research, two processes known as "donation" and "parcel subdivision" were the workflows focused on.

The requirements were primarily acquired using observation and semi-structured interviews, including open-ended questions. Secondary data in the form of reports and desktop research also informed the process. For the primary data, 15 actors involved in the updating of land information were identified and interviewed. The number includes seven right holders from the Musanze District Land Bureau (DLB) and eight staff working for one of the institutions participating in the updating of land information. Although the number of participants may appear limited, the data gathered were descriptively rich, would enable modeling of the processes, and at any rate were not intended for quantitative purposes. The specific aspects dealt with in the interviews included the following: the actors involved in the existing process, tasks performed, locations where transactions take place and number of visits involved, required documents and payments for the request to be processed, actors' interactions with technologies including the digital land register and what type of access they have, conditions to have access to the system, and quality checks performed. For parcel subdivision workflow, information was collected on the surveying tools used to take new measurements and the accuracy and required skills to manipulate them. All data gathered were synthesized to develop a general requirements set that could be used for both process modeling and validation phases of the research.

The modeling process was carried out in four steps: (1) primary and secondary data were analyzed and synthesized to develop "as is" or existing workflows for donation and parcel subdivision; (2) the modeled workflows were then analyzed against the general requirements set developed out of the interviews and direct observations: the strengths and any weaknesses of the process were identified and opportunities for redesign established; (3) from the obstacles and opportunities, more specific design requirements were established: the requirements were organized to be used in the validation phase, with each requirement linked to a specific obstacle; and (4) based on the defined design requirements an improved or "to be" workflow for updating land information was designed and presented using UML activity diagrams.

The validation process compared the "as is" model to the "to be" workflow: adherence to the design requirements was checked. Each requirement was assessed and allocated one of four options: (1) validated by the "to be" workflow, (2) partially validated for the "to be" workflow, (3) not validated for the "to be" workflow, and (4) already valid for both the "to be" and "as is" workflows. A limitation of the validation process was that the workflows were not presented and assessed by citizen user groups, although interaction and discussion with experts in the existing Rwandan land administration processes further informed the validation process.

Results and Discussion

First, the two "as is" workflows are presented (donation and parcel subdivision). Second, the specific design requirements established for each workflow are presented. Third, the "to be" workflow combining parcel subdivision and donation is justified and presented. Finally, the results from the validation are presented.

Donation (as is) is a change in land information where rights are transferred as a gift and commences with application. *Application* occurs when the donor goes to the DLB to ask for application requirements. He or she compiles the required documents and pays the required fees (Table 9.1). The application is taken to the LAIS professional at the DLB to check if all documents are complete and valid. "He or she" prepares a transfer agreement to be signed by the donor (with his or her spouse), the receiver, and four witnesses (two from each side). The application is sent to the district land officer (DLO),

TABLE 9.1

"As is" Collected Information on Parcel Donation

Element	Existing Situation
Four offices to enter	Sector office Bank DLB District reception
Four times to come to the district	Ask for information on rights transfer Sign transfer agreement Submit the application Collect land certificate
Eleven required documents	Copy of the donor's and receiver's identity card Donor marital status certificate Donation acceptance letter Filled application form Original land certificate Payment slip for transfer, notification, new land certificate, and cancelling the existing land certificate Notified donation agreement
Five required payments	20,000 RWF for transfer fees 2,000 RWF for notification 1,500 RWF for lease contract cancellation 1,500 RWF for new lease contract 500 RWF for marital status certificate
Three actors access the land information	Access to the land information of their province: PLR (view) LAIS professional at the ORLT (edit) DRLT (edit)

Source: Biraro, M., *Land Information Updating: Assessment and Options for Rwanda*, master of sciences, University of Twente, Enschede, The Netherlands, 2014.

acting as land notary, to notify the transfer agreement. The donor collects the application file with a notified transfer agreement and takes it to the district reception. The receptionist brings back the file to the district LAIS professional, who takes it to the Office of the Registrar of Land Titles (ORLT), at province level, for processing. *Processing* then commences at the ORLT. The application is received and manually recorded by the assistant of deputy registrar of land titles (DRLT). He or she takes the application to the professional in charge of land registration (PLR), who checks the completeness and validity of the documents. He or she also verifies, in the digital land register, if the donor is the real and only (known) right holder. If everything is correct, the application is sent to the LAIS professional (at the ORLT) for processing. If something is wrong, the application is taken to the DRLT, who approves its rejection. For the accepted applications, the LAIS professional scans and uploads the documents into the LAIS. The name of the receiver is added in the database, and the application is electronically sent to the DRLT for approval. This checks if the transfer was done according to the provided documents. If everything is correct, he or she approves it and the change is saved into the database. The LAIS professional prints and seals the new land certificate and takes it to the DRLT assistant. The district LAIS professional collects the printed land certificate and the rejected applications and takes them back to the DLB. *Issuance* commences with the district archivist receiving and manually recording files from the ORLT. After the given time, the receiver comes back, at the DLB, to collect the land certificate as the new right holder. If the certificate is available, he or she gets it. If the application was rejected, he or she gets an explanation of what to do so that the application can be processed. If there is no feedback yet, he or she gets a day that he or she could come back.

The workflow for parcel donation involves six actors, plus the donor and the receiver (Figure 9.1). The DLRT's assistant and the district archivist were omitted to simplify the activity diagram. The grouped columns represent offices where the application passes to be processed. The arrows between the first column and all other columns represent the times the applicant comes to the district. The grouped arrows are interactions done on the same day.

Parcel subdivision (as is) is a change in land information where a parcel is split into two or more parts. Where the split is followed by another change in land information (right transfer or land use change), the right holder follows two separate processes. The workflow is illustrated in four main steps. Application commences when the right holder goes to the DLB to ask for requirements. He or she compiles the required documents and pays the required fees (Table 9.2). The district LAIS professional receives and checks the application file to see its validity and completeness. If everything is correct, the file is given to the district surveyor, who makes an appointment with the right holder for field measurements. The right holder is in charge of transport of the surveyor when he or she comes to take measurements. *Surveying* is undertaken by the district surveyor, with a handheld global

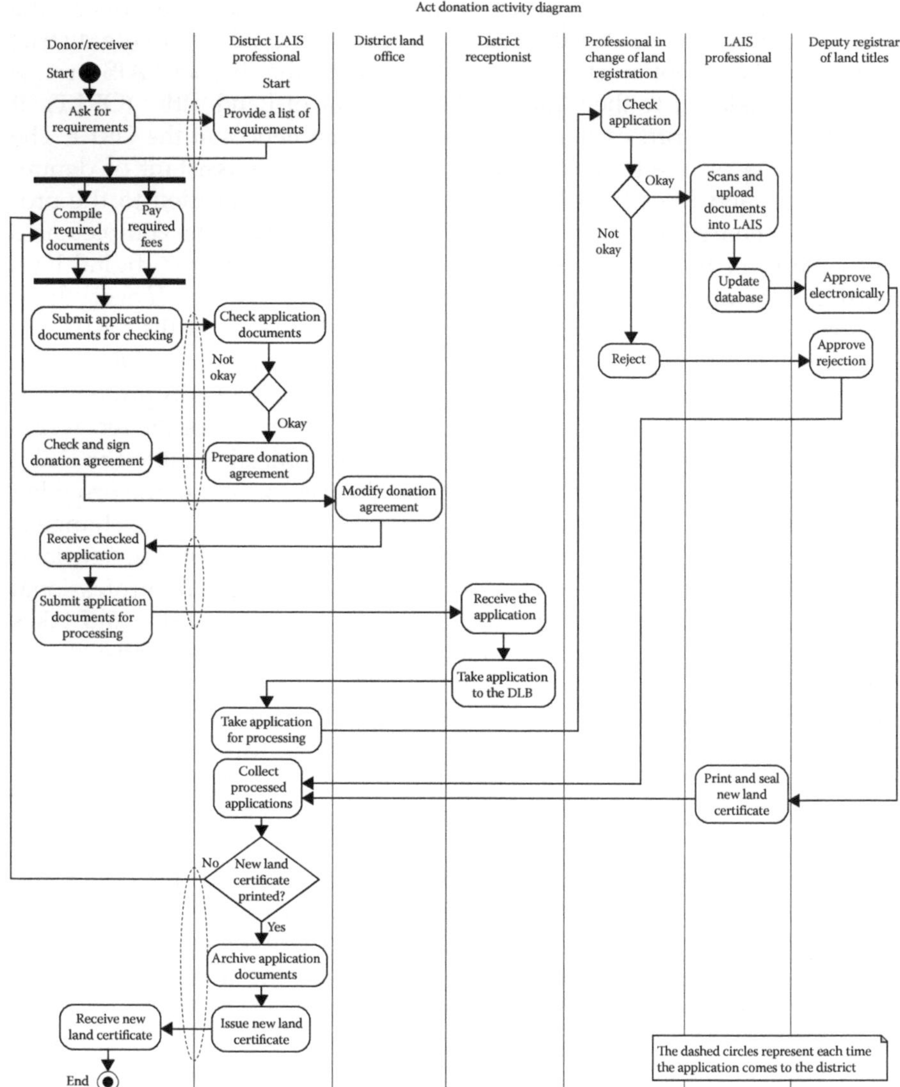

FIGURE 9.1
"As is" workflow for donation. (From Biraro, M., *Land Information Updating: Assessment and Options for Rwanda*, master of sciences, University of Twente, Enschede, The Netherlands, 2014.)

positioning system (GPS), who measures the new boundary that splits the parcel into two or more parts. The accuracy of the GPS is approximately 2 to 3 m, depending on conditions. Once in office, the district surveyor prepares the cadastral plans equal to the number of parcels to be created. He or she then takes the application documents to the DLO, who signs and stamps the cadastral plans. The district surveyor keeps the application file until the

TABLE 9.2

Collected Information on "As is" Parcel Subdivision

Element	Existing Situation
Four offices to enter	Sector office Bank DLB District reception
Five times to come to the district	Ask for information on rights transfer Submit the application for checking Submit the application for processing Collect land certificate
Surveying tools	Handheld GPS Printed aerial image
Skills	Basic surveying skills
Accuracy	3 m (For the GPS)
Nine required documents	Copy of the right holders' identity card Marital status certificate Application letter Filled application form Original land certificate Payment slip for cadastral plan, new land certificates, and cancelling the existing land certificate Cadastral plan for each of the parcels to be created
Four required payments	10,000 RWF for each cadastral plan to be produced 2,500 RWF for lease contract cancellation 2,500 RWF for new lease contract 500 RWF for marital status certificate
Four actors access the land information	Access to the land information of their province: PLR (view) LAIS professional at the ORLT (edit) GIS professional (edit) DRLT (edit)

Source: Biraro, M., *Land Information Updating: Assessment and Options for Rwanda*, master of sciences, University of Twente, Enschede, The Netherlands, 2014.

applicant comes to take it to the district reception. The receptionist brings back the application to the district LAIS professional, who takes it to the ORLT for processing. Processing is commenced by the assistant of the DRLT, who manually records the received application and takes it to the PLR. The PLR checks and verifies, in the digital land register, if the applicant is the real and only right holder. If everything is correct, the application is sent to the geographic information system (GIS) professional for processing. If there is something wrong, the application is taken to the DRLT, who approves its rejection. The GIS professional scans and uploads the documents of accepted applications into the LAIS. Using ArcGIS, he or she overlays the new measurements on the existing data sets. If they match, the parcel is split based on the coordinate points of the new boundary. New unique parcel identifiers are given to the created parcels. He or she prepares a report and a map explaining the change and uploads them into the LAIS. The application is digitally sent to the DRLT for approval. The LAIS professional prints and seals land certificates that he or she takes to the DRLT assistant. In case there is a mismatch between the new and the existing measurements, a field checking is done by two grassroots surveyors operating at the ORLT at province level. They contact the applicant to fix an appointment for this activity. With a printed map on which there are both new and initial measurements, they perform observations by discussing with the applicant and neighbors (if any). A field report is explained to the applicant and the present neighbors who sign if they agree with it. The report is given to the GIS professional, who processes the request if the problem was resolved. If not, the final decision on the application file is made by the DRLT. The district LAIS professional collects all the processed applications from the DRLT assistant and takes them back to the DLB. Issuance is commenced by the district archivist who manually records all the applications received from the ORLT. After the given time, the applicant comes back to collect the new land certificates, all on his or her names. If the new certificates are available, he or she gets them. If the application was rejected, he or she receives an explanation of what to do. If there is no feedback yet, he or she gets a day to come back.

The workflow for parcel subdivision involves eight actors, plus the right holder (Figure 9.2). The DLRT assistant, district archivist, and grassroots surveyors were omitted to simplify the diagram. The grouped columns represent offices where the application passes to be processed. The arrows between the first and all other columns represent the times that the applicant comes to the district. The grouped arrows are interactions done on the same day.

The aforementioned workflows have many positive points to highlight like the LAIS now operational in all the five ORLTs at province level, right holders who are aware of reporting some of the land information changes linked to rights transfer, and people who understand the importance of having a land certificate as they exchange the land certificate if they do not go to report the change (Biraro 2014). They evidence that the maintenance phase is being strengthened. The remaining obstacles include the long travel distance to the DLB, the need to attend several times at places where the change is being reported due to the limited use of ICT in service delivery, the long updating process due to an analogue approach still in use and to the many actors involved in the process, the many required physical documents for which some are found in different places, and the high registration fees added to other costs spent on compiling the required documents (Biraro 2014).

Design requirements were defined and used in the design (Table 9.3). This would assist in proposing a way to handle the remaining obstacles in land information updating, and validating the subsequent designs. They include what the right holders want the system to do for them and what the system needs to have to perform its functions well (Norman 1996).

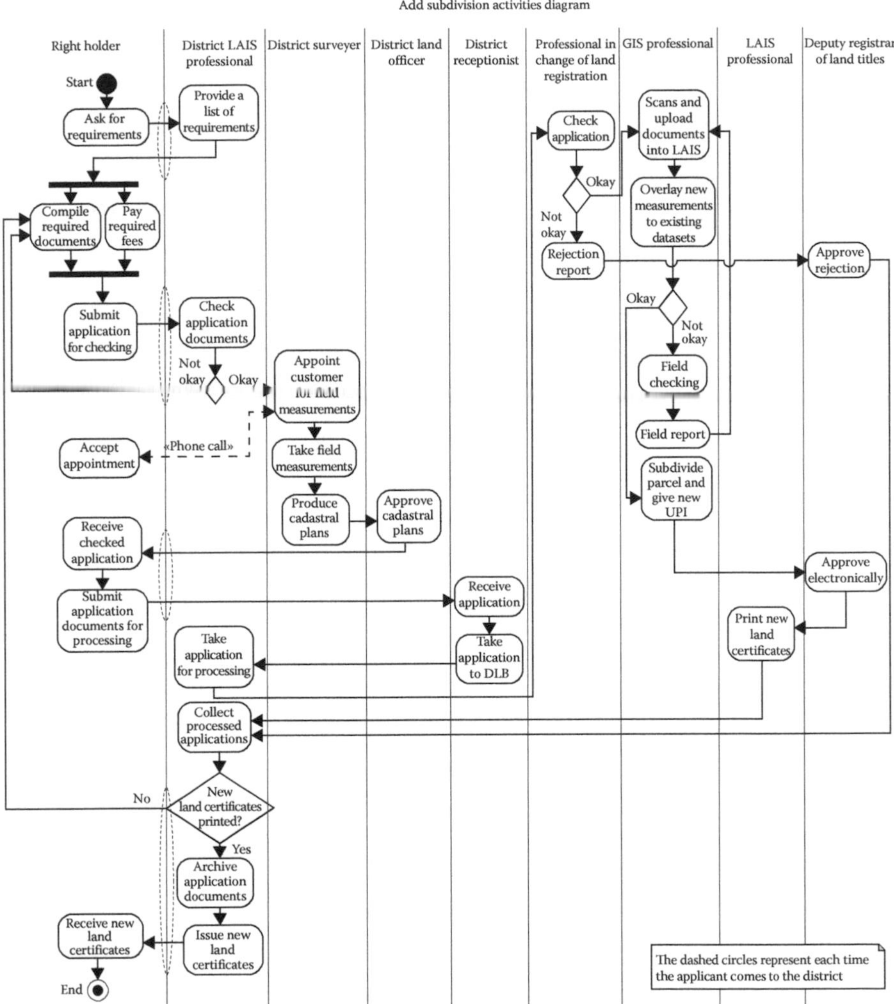

FIGURE 9.2
"As is" workflow for parcel subdivision. (From Biraro, M., *Land Information Updating: Assessment and Options for Rwanda,* master of sciences, University of Twente, Enschede, The Netherlands, 2014.)

The workflow redesigns needed to meet the defined requirements. Analysis of the requirements revealed that they could be best met via one comprehensive workflow consisting of interconnected database systems (Biraro 2014). The workflow connection would reduce the number of submitted documents and save time and money for the right holder. Consequently, Biraro (2014) proposed that the LAIS should be connected to the (1) National Identification Project, which would contain information about the marital status of every citizen with an identity card and a unique identification

TABLE 9.3
Design Requirements

Obstacle	Design Requirement
Coming several times to the DLB.	Easy way of getting information at local offices or online.
Weak use of ICT in service delivery.	Come to the DLB a few times. Contact the customer via e-mail or phone when his or her presence is needed.
Long process for updating land information.	One-stop center for land services. Processes can be combined where possible. Involved actors should be reduced. Reduce the use of physical documents.
Many required documents.	Reduce submitted documents through systems interconnection.
High registration fees.	Determine the required fees based on the capacity of people to pay. Payment should be done after confirming that the enquired service will be provided.
Lack of access to land information.	Actors involved in the updating process should access the land information. Control access to the land information database.
Long surveying process.	One actor should do the surveying and data processing and correct his or her own mistakes.
Accuracy mismatching.	Use the initial data sets as a starting point for new measurements to improve the accuracy and reduce the process time. Simple surveying tools, easy to manipulate. Flexible accuracy depending on the purpose.

Source: Biraro, M., *Land Information Updating: Assessment and Options for Rwanda*, master of sciences, University of Twente, Enschede, The Netherlands, 2014.

number (from 16 years old); (2) court with a database on land-related juridical cases (this would allow checking if there is no pending case on the parcel for which the information is to be changed); (3) Rwanda Revenue Authority, which would inform the new right holder about land taxes, if they are paid or not, or how much he or she will be paying in the future; and (4) banks, with e-payment services, and the DLB would know if the applicant transferred the required fees on the district account.

The subsequent design of the proposed workflow, based on this connectivity, combined the processes of donation and parcel subdivision. According to Biraro (2014), rights transfer or land use change is mostly the reason behind a parcel subdivision. So, why not have a combined workflow instead of having two separate processes? The right holder would apply only once, and two land certificates would be produced with different names. In addition, with the proposed workflow two possibilities are offered, *online* and *physical* applications, taking into account the current technology availability and capacity within the district.

Combined workflow (to be): online includes application, field survey, processing, rights transfer, and issuance. Application commences when the donor is given an account by the IT manager to be able to apply online. The donor logs in, completes the application form, and uploads the scanned land certificate. After submission, he or she gets an application identification number as an approval that the documents are received. The DLO, after logging in, sees all received applications. By using the unique parcel identifier (UPI), he or she retrieves information about the concerned parcel to verify if the donor is the owner and the only owner and if there is no mortgage, dispute, or unpaid taxes. When everything is cleared, the application is sent to the district surveyor for field measurements. If there is a problem, the DLO writes a rejection report and sends it to the donor. *Field survey* commences with the district surveyor locating the parcel of the received application using the initial data set and the UPI. He or she prepares a field map where the initial parcel boundaries are overlaid on an aerial/satellite image. An appointment is fixed with the donor through e-mail. In the field, guided by the donor and the receiver, the district surveyor confirms the shape of the parcel using the printed field map. He or she surveys the new boundaries using a GPS or any other surveying approach depending on the purpose. A field report with the used map, showing where the measurements were taken, is written and explained to the donor and the receiver who have to approve it. Processing commences in the office. The district surveyor scans and uploads the field report into the LAIS and starts to process the application. The new measurements are overlaid on the existing data sets. If there is something wrong, he or she goes back to the field to correct the mistakes. But if there is no mistake, he or she splits the parcel, gives new UPIs, and produces cadastral plan for each of the created parcels. He or she digitally sends the cadastral plans to the DLO to do the transfer. *Right transfer* occurs when the DLO prepares the transfer agreement to be sent to the donor together with the cadastral plans via e-mail and specifies when to come for signature. Both the donor and the receiver check the documents. If they agree, they respond to the invitation. If not, comments are sent to the DLO. On the agreed day, the receiver pays the required fees, using e-payment. The two parties (with their spouses, if any) sign the transfer agreement, and the original land certificate is submitted to be canceled. The DLO notifies the signed transfer agreement, scans it, and uploads it into the LAIS. The application is digitally sent to the LAIS professional to add the change in the database. The DRLT checks the change made and approves the transfer if everything is correct. The database is updated, and new land certificates are ready for issuance. Issuance happens when the DLO checks the land certificates ready for issuance and notifies their owners. The land certificates are printed when their owners come to collect them.

Combined workflow (to be): physical application is used for those who do not have Internet access. It should be noted that processing and issuance occur in the same fashion as the online application and are therefore not included in this description. Application commences when the donor goes to the leader of the village (the smallest administrative unit) to get a list of requirements and an application form. As the right holder, he or she fills the form and attaches the original land certificate. The application documents are taken to the PLR as a front desk officer at the DLB. This one checks, in the digital land register, the owner, if the parcel is not mortgaged, and if no juridical case is pending or there are no unpaid land taxes. If everything is cleared, the PLR scans and uploads the application documents into the LAIS and gives an application identification number that is automatically generated by the system. He or she archives the physical documents and sends the digital copy to the DLO. If there is a problem in the application, the donor receives an explanation of what is wrong. The DLO also checks the application before sending it to the district surveyor for measurements. If there is a problem, the DLO writes a rejection report and sends it to the PLR, who informs the donor through a phone call. Field survey is undertaken by the district surveyor: the UPI helps to locate the parcel of the received application. He or she prepares a field map where the initial parcel boundaries are overlaid on an aerial/satellite image and fixes an appointment with the donor through a phone call. The field measurements are similar to what is done during the online application. Right transfer happens when the PLR contacts the donor and the receiver to sign the transfer agreement. On the agreed day, the DLO prepares the transfer agreement with the cadastral plan and shows them to both parties. They check the documents, and if they agree with the content the receiver pays the required fees to the DLO. The two parties (and their spouses, if any) sign the transfer agreement.

Validation revealed that requirements were validated, partially validated, or not validated in the "to be" workflows. Additionally, some were found to be already validated in the existing or "as is" workflow.

A total of 10 requirements were identified as being validated: (1) only one process is to be followed in case the transfer is for a part of the parcel; (2) information on updating will be available both online and at village level (closer to the people) instead of only being acquired at district level; (3) an applicant needs to come to the DLB only two times (with the support of an online application) or three times (physical application) instead of eight times (dashed circles in Figures 9.1 and 9.2); (4) the number of involved actors can be reduced from 14 actors (8 during parcel subdivision and 6 during donation) to 5 or 6 actors in online or physical application, respectively; (5) only the professional in land registration (physical application) and the DLO (online application) will use the physical documents while almost all the actors are using them in the existing workflow (6 out of 8 in parcel subdivision and 5 out of 6 in donation); (6) due to system interconnection, the submitted physical documents can be reduced from 20 documents (11 for donation, Table 9.1, and 9 for parcel subdivision, Table 9.2) to 2 documents; (7) the payment can be done after the application is accepted while, in the existing situation, it was done before the application is submitted; (8) actors at the DLB involved in the updating process can now access the land information; (9) surveying and processing activities can be done by 1 actor while, in the existing situation, it was involving 4 actors (district surveyor, GIS professional, and two grassroots surveyors); (10) instead of starting from scratch during the surveying, the initial data sets can be used as a starting point for new measurements.

Meanwhile, two requirements can be identified as being partially validated: (1) only applicants with e-mail/phone can be contacted; those without e-mail/phone will still be coming to check the status of the application. (2) The land services need to be offered at two separate offices, district and province levels. Meanwhile, one requirement is not validated: as cost analysis was not covered under this research, determination of required fees based on the capacity of people to pay was not met. Finally, three requirements were found to be already valid: (1) access to land information database will be controlled; (2) the accuracy of surveying depends on the purpose; and (3) simple surveying tools, easy to manipulate, will be used.

The validation suggests that the proposed (to be) workflow better meets the requirements than the two separated existing (as is) workflows. Most of the obstacles can be handled; thus, the right holders are expected to be more motivated to report changes. However, the design requirements that could be only partially validated or not validated require further research.

Conclusion

Updated land information is critical for a well-functioning land administration system. Even though the first registration is systematic, the updating is sporadic. Right holders initiate the process by reporting the change. All obstacles standing in the way of the updating process should be minimized or removed to get the full collaboration of the right holders. If this is not organized, they may opt for informality and the system will finally lose its value. The updating workflows should be simple and customer oriented to facilitate their users. The workflow design should respect requirements allowing identifying and handling those obstacles in the updating process to ensure that changes on the ground are getting in the land register.

This chapter discussed the use of workflow management to improve the updating process in land administration. Many authors agree on the necessity of updating the land information, and how efficient the process should be. However, less was said on how these processes should be designed to speed up the reporting of changes in land information. By using Rwanda as a case study, a workflow management system designed with UML was used to analyze the existing updating system and also to design improved processes.

The methodology proved to be useful as it allowed identification of obstacles in the workflow and the definition of requirements. Improvements in the proposed workflow were validated such as combinations of processes instead of following two separate workflows when two changes are on a same parcel, fewer actors, and less use of papers in the process. All of these have an impact on the speed of the updating process. The workflow based on UML diagrams was beneficial in describing, step by step, the updating process. Specific to Rwanda, future work should determine whether the proposed designs are suitable among right holders. From a theoretical perspective, work could focus on developing measures for ensuring that land information–updating processes receive equal attention, or are even entirely integrated into the design of initial registration programs.

References

Aguilar-Savén, R. S. (2004). Business process modelling: review and framework. *International Journal of Production Economics, 90*(2), 129–149.

Aversano, L. et al. (2002). Business process reengineering and workflow automation: a technology transfer experience. *Journal of Systems and Software, 63*, 29–44.

Binns, B. O., and Peter, F. D. (1995). *Cadastral Surveys and Records of Rights in Land: An FAO Land Tenure Study* (e-book), accessed November 28, 2013, http://www.fao.org/DOCREP/006/V4860E/V4860E00.HTM.

Biraro, M. (2014). *Land Information Updating: Assessment and Options for Rwanda.* (Master of sciences), University of Twente, Enschede, The Netherlands.

Chimhamhiwa, D. et al. (2011). Measuring quality performance of cadastral survey and deeds registration work processes. *Land Use Policy, 28*(1), 38–46.

Chimhamhiwa, D. et al. (2009). Towards a framework for measuring end to end performance of land administration business processes: a case study. *Computers, Environment and Urban Systems, 33*(4), 293–301.

Deininger, K. et al. (2012). *The Land Governance Assessment Framework: Identifying and Monitoring Good Practice in the Land Sector*. Washington, DC: the World Bank.

Enemark, S. (2004). *Building Land Information Policies.* Paper presented at the Special Forum on Building Land Information Policies in the Americas, Aguascalientes, Mexico.

Enemark, S. et al. (2014). *Fit-For-Purpose Land Administration.* Joint publication by FIG and World Bank, Copenhagen, Denmark.

FAO. (2012). *Voluntary Guidelines on the Responsible Governance of Tenure of Land, Fisheries and Forests in the Context of National Food Security*. Rome, Italy: FAO.

FIG. (1995). Statement on the cadastre, accessed August 7, 2013, http://www.fig.net/commission7/reports/cadastre/statement_on_cadastre.html.

Glassey, O. (2008). A case study on process modelling—three questions and three techniques. *Decision Support Systems, 44*, 842–853.

Heloisa, M. S., and Mitchell, M. T. (1996). Workflow technology-based monitoring and control for business process and project management. *International Journal of Project Management, 14*, 373–378.

Henssen, J. (2010). *Land Registration and Cadastre Systems: Principles and Related Issues.* Technische Universität München, München, Germany.

Jing, Y. et al. (2013). *'Up-to-Date' in Land Administration: Setting the Record Straight.* Paper presented at the Environment for Sustainability—FIG Working Week 2013, Abuja, Nigeria.

Larsson, G. (1991). *Land Registration and Cadastral Systems: Tools for Land Information and Management*. New York: Longman Scientific and Technical.

Lemmen, C. (2012). *A Domain Model for Land Administration.* (PhD), Delft University of Technology, Delft, The Netherlands.

Magis, M., and Zevenbergen, J. (2014). *Towards Sustainable Land Administration Systems: Designing for Long-Term Value Creation.* Paper presented at the FIG Congress: Engaging the Challenges, Kuala Lumpur, Malaysia.

Mentzas, G. et al. (2001). Modelling business processes with workflow systems: an evaluation of alternative approaches. *International Journal of Information Management*, *21*(2), 123–135.

Norman, R. J. (1996). *Object-Oriented Systems Analysis and Design*. London, United Kingdom: Prentice Hall.

Scheu, M. et al. (2000). Incremental update and upgrade of spatial data. *Zeitschrift fur Vermessungswesen*, *4*, 115–120.

Todorovski, D., and Lemmen, C. H. J. (2007). Analysis of user requirements: the first step towards strategic integration of surveying and cadastral services. FIG Working Week 2007: Strategic Integration of Surveying Services, Hong Kong, China.

UNECE. (1996). *Land Administration Guidelines with Special Reference to Countries in Transition*. New York; Geneva, Switzerland: UNECE.

Williamson, I. et al. (2010). *Land Administration for Sustainable Development*. Redlands, CA: ESRI.

Zevenbergen, J. A. (2002). *Systems of Land Registration: Aspects and Effects.* Delft, The Netherlands: Netherlands Geodetic Commission.

Zevenbergen, J. A. et al. (2007). *Real Property Transactions: Procedures, Transaction Costs and Models*. Amsterdam, The Netherlands: IOS Press.

Zur Muehlen, M., and Indulska, M. (2010). Modeling languages for business processes and business rules: a representational analysis. *Information Systems*, *35*(4), 379–390.

10

Toward Fit-for-Purpose Land Consolidation

Rohan M. Bennett, Fikerte A. Yimer, and Christiaan Lemmen

CONTENTS

Introduction

Land consolidation is the process of reallocating, and sometimes spatially redesigning, rural landownership patterns to create more rational landholdings (Pašakarnis and Maliene 2010). The conventional process finds its origins in Western European countries where it is practiced at scale, and was variously applied over the last 200 years with the aim of improving agricultural production and social conditions (Crecente et al. 2002; Hartvigsen 2005). Consequently, conventional land consolidation methods are heavily imbued with understandings, norms, and technologies developed in Western Europe.

Land consolidation appears to have application beyond Western Europe, particularly in sub-Saharan African countries, where population growth and urban migration potentially conspire to undermine food security. In these contexts, much emphasis is placed on using new geospatial tools to fast track the creation of functioning land administration systems. Rural communities, especially smallholder farmers, play an important role in the sustainable management of natural resources that are critical for global food security (FAO 2013). However, high parcel fragmentation often undermines food security in these contexts. Where land administration systems are functioning, they could support delivery of food security through the enablement of responsible land consolidation activities.

Application of land consolidation in sub-Saharan African countries remains relatively unexplored. It is uncertain whether land consolidation can, or should, be applied in such settings. Moreover, if land consolidation has relevance, it is not clear whether conventional methods could be applied directly, whether adaptations and modifications would be required, and what the nature of those adaptations might be. To this end, and in line with the design orientation of this section of the book, this chapter seeks to determine the relevance of land consolidation for a specific sub-Saharan African context and to illustrate how it could be applied. Put simply, the chapter aims to develop a fit-for-purpose land consolidation procedure, one that aligns with the FIG's (2014) fit-for-purpose land administration ideals.

Ethiopia is considered to be food insecure. The majority of the population comprises smallholder farmers (~90%). At least 35% of the population is undernourished (WFP 2014a,b). Despite this, the country has experienced high economic development rates since the late 1990s. Its large population and geopolitical importance mean that it will continue to be a touchstone for evaluating economic development in the region (Bennett and Alemie 2015). Land consolidation appears to have relevance for parts of the country: population growth fuels unsustainable parcel fragmentation among rural communities (Gebreselassie 2006; Teshome et al. 2014). Land consolidation is already allowed on a voluntary base; however, as yet there is little evidence of any significant consolidation activities. However, it could be highly supportive of the Ethiopian government's Growth and Transformation Program, which aims to double the productivity of smallholder farmers while keeping large-scale agricultural lands for the production of cash or export items (FDRE 2010). If undertaken in a responsible fashion, a systematic approach to land consolidation could improve the productivity of smallholders and improve food security more generally.

This chapter investigates how conventional land consolidation concepts might be adapted to suit the Amhara region of Ethiopia. The resulting showcase demonstrates how fit-for-purpose land reallocation* can be achieved when digital cadastral data are available and geographic information system (GIS) tools are applied. The research begins from conventional land consolidation methods and develops them into something new and innovative: processes from the Netherlands are modified for the Amhara region in Ethiopia. The importance of understanding and working with local stakeholders is revealed. This work is considered the first of its kind: there exists no prior research on how land consolidation concepts might be applied in Ethiopia. It is hoped that the results contribute to the domain of land consolidation, particularly agricultural reallocation plans for Ethiopia's rural settings. The remainder of this chapter is structured as follows. First, a background on land consolidation, land reallocation, land banking, and the Ethiopian land tenure system provides theoretical context. Second, the socio-technical design

* Considered synonymous with "reallotment" in this chapter.

approach and a selected case area are described. Third, results from the various phases of the research are articulated. Fourth, the implications for the results and their application are discussed. The chapter concludes with a summary of key findings and an articulation of future research directions.

Theoretical Perspective

Conventional land consolidation, performed at scale, is a product of Western European countries and was more recently applied in Central and Eastern Europe (Sabates-Wheeler 2002; Sklenicka 2006; Van Dijk 2003) and China (Wu et al. 2005). Land consolidation is described by Vitikainen (2004) as follows:

> a comprehensive reallocation procedure of a rural area consisting of fragmented agricultural or forest holdings or their part.

However, more generalized definitions (Pašakarnis and Maliene 2010) exist. The concept overlaps with both agrarian spatial planning and land readjustment. The first land consolidation initiatives emerged in Denmark in the 1750s: the establishment of private land to free people from obligations to nobles was the aim (FAO 2003; Hartvigsen 2005). Most programs have a legal basis: the process can be compulsory or voluntary. In the Netherlands, the first law for land consolidation was issued in 1924 and revised in 1938. However, more substantial consolidation commenced after World War II in the framework of a land development project integrated with infrastructure and landscape development objectives (Bergh 2004; Thomas 2006a; Van den Noort 1987). A key aim was to improve agricultural productivity and to minimize production cost. The process continues in the contemporary era (Kadaster 2012; Kool 2013; Rienks et al. 2009; Van den Brink 2004). Beyond these aims, land consolidation has also been used as an instrument for rural development; enhancing rural life quality; and improvement of infrastructure, employment, housing, environmental protection, agricultural production, village development, nature preservation, soil conservation, and outdoor recreation (Damen 2004; Lemmen and Sonnenberg 1986; Mihara 1996; Thomas 2006a,b). Overall, it is argued as a means for sustainable improvement of livelihoods in rural areas (Buis and Vingerhoeds 1996; Thomas 2004). However, others reveal the potential negative impacts on social and environmental conditions (Bonfanti et al. 1997; Zhengfeng and Wei 2007). Others show that land consolidation, at least in the context of south Asia, does combat parcel fragmentation in countries where underlying structural issues are not addressed first (Niroula and Thapa 2005). At any rate, regardless of where it has been applied, land consolidation involves or interacts with five elements: land reallocation, land banking, spatial technologies, land tenure systems, and fragmentation. These are now discussed.

An important subconcept, if not the key component, relating to land consolidation is *land reallocation* (or reallotment). Rosman (2012) described land reallocation as the driving force behind land consolidation and subsequent land use improvements. The reallocation process can minimize transportation expenses and labor cost to landowners or users: different owners can be allocated a more advantageous position by locating farmlands near farmhouses (Buis and Vingerhoeds 1996). The process of reallocation in the Netherlands was designed, and is still designed, to restructure land parcels and to ensure more efficient agricultural activity: large parcels of land, improved form or shape, and a reduction of distance between the farmhouse and parcels result in a reduction of costs for farmers. If implemented correctly, more land becomes available for cultivation: higher yields for individual farmers result (Buis and Vingerhoeds 1996). If the aim of land reallocation is to reduce fragmentation and to merge uneconomically shaped parcels, voluntary land exchange is cited as the simplest and fastest land consolidation measure. Demetriou et al. (2012) explain that reallocation involves two processes: (1) land redistribution, or restructuring of land tenures into larger blocks, and (2) land portioning, or re-parceling of those blocks into smaller spaces. The outcome of the process is a context-specific reallocation plan (Buis and Vingerhoeds 1996).

Another related concept, *land banking*, is as follows:

> the structural acquisition and temporary management of land in rural areas by an impartial State agency, with the purpose to redistribute and/or lease out this land … for … a general public interest (Damen 2004).

Land banks help to transform land in terms of the following: (1) place—one can sell in one location and buy in another location, (2) quantity—small parcels can be joined to larger parcels, and (3) quality—one can sell or give an amount of lower quality land in return for a smaller quantity of better quality land and vice versa. The land bank "combines" the land of many owners and can help make much more land combinations temporarily available (Jansen and Wubbe 2012). Land banks help to solve problems during a land consolidation project: they facilitate and balance allocated values and allocation claims (Jansen et al. 2010). Indeed, Damen (2004) concludes that successful land consolidation in the Netherlands was dependent on land banking.

Spatial technologies, including GIS, are now used heavily in land consolidation activities. Lemmen et al. (2012) and Demetriou et al. (2011) expand on the design process, its recent computerization, and the application of GIS. Many reallocation algorithms are developed in different countries, including Finland (Tenkanen 1987), Germany (Hupfeld 1971; Schrader 1971), the Netherlands (Kik 1971, 1979; Lemmen and Sonnenberg 1986), Spain (Touriño et al. 2003), and Turkey (Ayranci 2007; Cay et al. 2010). These approaches may be heuristic in nature, or based on optimization (e.g., distance reduction or plot concentration). Well-known algorithms embedded into design systems include the stepping stone algorithm (Kik 1971); Automation of the Reallocation Plan for land consolidation (De Vos 1982) and its more recent iteration TRANSFER (Rosman 2012); allocation and adjustment model (AVL) (Lemmen and Sonnenberg 1986); and transportation algorithm, mixed integer programming, and simplex method (Lemmen et al. 2012). In more recent times, these tools have been combined with participatory methodologies to enable on-site design processes, sometimes in real time (Louwsma 2013; LTO Nederland 2013).

To understand how conventional land consolidation might be adapted and applied in the Ethiopian rural context, the *land tenure system* needs consideration. The system can be considered through chronological lenses: before 1975, 1975 to 1995, and post 1995 (Nega et al. 2003; Teka et al. 2013). In the contemporary era (i.e., post 1995), all land is considered common property for the nation and people of the country: government considers that private landownership results in unequal landholdings, landlessness, and rural–urban migration by the rural poor (Hoben 2000; Nega et al. 2003). Land is not subject to sale or other means of exchange; however, Article 40 Sub Art 4 of the 1995 Ethiopian constitution states that

> Ethiopian peasants have the right to obtain land and protection against eviction.

Riddell and Rembold (2002) argue that the land tenure system makes it difficult to improve infrastructure and perform environmental protection for sustainable economic growth, social development, and sustainable rural livelihoods.

The majority of Ethiopian peasants depend on allocated plots for agricultural production to generate income for their subsistence (Belay 2010). *Land fragmentation* is considered a significant problem in some regions of the country (Nega et al. 2003; Teka et al. 2013): the chance for sustainable agricultural intensification is very much limited where peasant agriculture is characterized by small and spatially disparate plot holdings (FDRE 2003; Gebreselassie 2006). Indeed, Gebreselassie (2006) reveals that during the 2000 cropping season 87.4% of rural households operated with less than 2 ha. On average, the farms were spread across 2.3 plots that were badly shaped and nonadjacent. Only 50% of the minimum household income to sustain life is delivered through such farms. The increasing fragmentation leads to declining soil quality, fertility, cash income, and technology investment and also results in higher overheads (Gebreselassie 2006). Despite these drawbacks, farm fragmentation occurs only on a voluntary basis: via bequeaths, smallholders share plots with children. The federal government encourages land consolidation, but only on a voluntary small-scale basis. Meanwhile, the contemporary government objective to modernize agricultural production to deliver food self-sufficiency (FDRE 2010) adds complexity. The policy allows land acquisitions by foreign and domestic investors. These affect rural households and result in reduced cultivated land for smallholders (Alemu 2012). Deininger and Byerlee (2011) explain how the total land transferred to investors between 2004 and 2008 (1.2 Mha) was a significant amount and suggestive of substantial changes to the existing agrarian structure.

Methodology

The theoretical background reveals that conventional land consolidation, land reallocation, and land banking approaches are yet to be tested in the Ethiopian context. The four-phase methodology to support the design-oriented research included (1) conceptualization, (2) data collection and analysis, (3) design and development, and (4) testing and evaluation. It should be noted that although comprehensive land consolidation is a broad application as understood by Thomas (2004), time and resource limitations meant that the design was limited to a reallocation plan with distance reduction and maximization of concentration of lots being the key foci. This easier to implement approach, in terms of cost and time, was considered more suitable to the Amhara region, where economic constraints would be a significant barrier to infrastructure development. Additionally, only methodologies from the Netherlands were considered: Dutch approaches were already applied in other development contexts (e.g., Turkey) and were found to be adaptable.

Phase 1, conceptualization, focused on the establishment of a theoretical frame, a systematic literature review, and selection of a specific case study location. Prior to commencing design work, a specific case location was identified. Ethiopia's ecologically diverse Amhara region was selected (area = 179,752 km^2; population = 19M (est.)) (CSA 2012). The region is divided into highlands and lowlands and has three climatic zones (FDRE 2013). Agriculture is the predominant economic activity in the region: 87.4% of the households depend on agriculture for their subsistence. Crop production is the main income source of the households, followed by livestock production. The region is administratively structured into 10 zones, which are further divided into 128 woredas and 3100 kebeles. The region is one where rural land registration and certification has been carried out: the essential data on land use boundaries were already available. An area comprising 70 households was selected within the woreda of Fageta Lakoma. Farming practice in this woreda has largely been determined by shortage of land and prevalence of very small landholdings: land is highly fragmented, with many smallholders having four plots in different locations. The kebele of Tafoch Dambule was selected from this region for data collection and subsequent reallocation design.

Phase 2, data collection and analysis, included three main activities. First, base data on landholdings and perceptions on consolidation and reallocation were collected from the selected 70 smallholder farmers (at the household level) using interviews, questionnaires, and focus group discussions. A pre-prepared hard copy base map supported the participatory sessions. Data collection with the smallholders took 8 days: each farm required a 30-minute interview (Figure 10.1). In addition, perceptions from regional and federal government officials were recorded using semi-structured interviews. Participants were located in Addis Ababa and Bahir Dar and included the Federal Ministry of Agriculture, Regional Bureau of Environmental Protection and Land Administration, and Woreda Land and Environmental Protection Office. Second, capture of secondary data and creation of a geodatabase (using ArcGIS) for the case area was undertaken. The database included separate tables for people, users, and "owners" (or recognized holders or smallholders). The annotated hard copy map used in the field was also scanned and digitized. Other spatial data captured and incorporated into the geodatabase included cadastral parcels, existing plot holdings, boundaries, soil types, topographic information, and imagery from Ethiopian government offices and other online sources. Third, capture of knowledge from a focus group discussion with experts from the Netherlands Kadaster, in Apeldoorn, on Dutch land consolidation experiences and approaches was

FIGURE 10.1
Study area aerial view, data collection with smallholders, and teff crop.

undertaken: software selection and reallocation algorithms were informed by these discussions. A full account of the data capture procedure is available in the work by Yimer (2014).

Phase 3, design and development of the reallocation plan, used the ArcGIS geodatabase, spatial analysis, and adapted algorithms to perform a showcase reallocation. The approach consisted of three main steps and was inspired by the optimization work and technical procedures of Essadiki (2002), Essadiki and Ettarid (2002), and Essadiki et al. (2003). Additionally, the Dutch swapping approach for voluntary agricultural reallocation practice was also adapted (Louwsma 2013; LTO Nederland 2014). The first step involved transforming the preferences of smallholders, collected from field questionnaires, into criteria. Plots and farmhouses (smallholders could have more than one) were the central units of analysis. Each plot was then ranked based on the criteria and assigned a value, and a sum provided for each smallholder. The plots that could be reallocated were then identified (e.g., ones with no farmhouses, or plots at large distances from a farmhouses). The second step included performing the exchange of plots among smallholders. The third step involved mapping of the newly relocated smallholders' plots with different colors.

Phase 4, testing and evaluation, involved assessing a number of predetermined indicators. These included gathering overall perceptions of smallholders and government officials toward the land consolidation concept and calculating outputs resulting from the reallocation showcase. For the latter, the existing situation was compared against the showcase reallocation with respect to total numbers of plots per smallholder, total plot sizes and plot shapes (assessed by the area to perimeter ratio), distance between plots and houses for an individual smallholder, and concentration of plots (assess by area multiplied by distance).

Results

Results from phases 2 through 4 are now presented: phase 1 results were presented in the theoretical perspective and methodology sections. Results from phase 2 describe stakeholder perceptions. Phase 3 results quantify reallocation outputs from the design process. Phase 4 results illustrate the evaluation of the reallocation against the five selected criteria.

Results from the field surveys, interviews, and focus group discussions revealed the existing perceptions of smallholders regarding land consolidation generally. From the total of 70 sample households, all suggested that a shortage of land impedes adequate food production. A single household has on average 1.22 ha of land. In addition, most of the plots were fragmented and not adjacent to each other. One household may have on average 4.75 plots in different locations and travel an average of 1.5 km between plots. Many respondents suggested there were difficulties in earning extra income at local markets, although the specific reasons were not captured. Interestingly, only 31 of the households considered that land reallocation could improve agricultural productivity. The other 39 did not consider reallocation as a solution for the improvement of farm structure or agricultural productivity. The smallholders provided different reasons for not considering the reallocation plan as a solution. The main reason was variation of soil quality between plots: smallholders want to keep plots in different places to ensure a variety of soils and subsequent crop diversity. The second reason was the threat of natural disaster: when farm plots are spatially spread, vulnerability to natural disaster (i.e., flood and wind) is reduced as is the likelihood that all crops are destroyed at the same time.

Results from the meetings with various Ethiopian government officials revealed the existing governmental perceptions on land consolidation. The federal level encourages land consolidation on a voluntary basis and is aware of the probable need for a more widespread large-scale reallocation program in the future. Additionally, agricultural land banks exist in both rural and urban areas in Ethiopia; however, they are only used to support large-scale land investments and capture of lands from deceased persons. They do not support land consolidation activities like those found in European countries. Currently, land reallocation is constitutionally beyond the jurisdiction of the federal government: for land consolidation to be undertaken beyond a voluntary basis, changes to, and the meaningful implementation of, a national land policy is required. At the worede and regional level, there was agreement that land shortage represented a major issue for smallholder agricultural production. Moreover, it was agreed that fragmentation was increasing. Although not averse to the consolidation concept, the representatives stressed that there were no plans to impose any form of reallocation activity: chemical fertilizer and resettlement programs were preferred in the short term.

The focus group discussions with the Netherlands Kadaster revealed potential lessons for land consolidation application in Ethiopia. Two Dutch approaches appeared to have potential: (1) voluntary reallocation and (2) formal land reallocation. In voluntary reallocation, the process was based on smallholders' wishes. Interactive sessions with individual smallholders are undertaken to design new allocation plans. These are followed by an interactive group session: possibilities of reallocation and the new scenarios are discussed. Thus, the smallholders prepare the reallocation plan together with the experts. Meanwhile, in formal reallocation every landholder must participate in the reallocation, since the aim is to improve the overall situation of the landholders (the farm size) and/or to keep the same status of the landholder. Hence, if the landholders were displaced from their original plots, the government gives them comparable lands elsewhere, using the land bank. The experts also provided lessons on software selection and application. Standard GIS tools including MapInfo, ArcGIS, ArcView, and Geo-Media, together with MS Access databases, were suitable for voluntary reallocation. More complex and large-scale formal reallocation activities made use of TRANSFER.

Regarding the actual design of the reallocation showcase, the procedure outlined in the methodology section was followed: (1) determining which plots could be reallocated based on smallholder preferences (i.e., whether plots had farmhouses on them, households having multiple plots, plot adjacency for individual smallholders, and soil quality), (2) exchanging plots among the smallholders (the swapping approach), and (3) identifying and mapping the reallocated smallholder plots with different shades (Figures 10.2 and 10.3). Based on the criteria, 30 of the 70 households were included in the reallocation process.

Regarding the testing and evaluation phase, the showcase was evaluated based on evaluation factors calculated before and after the reallocation. These included total designed plot size, total number of reallocated plots per smallholder (note: neighboring plots were considered a single plot), concentration of designed plots (in terms of area multiplied by distance from the smallholder houses), designed plot shape (determined by the area to perimeter ratio), and closeness of the designed plots to the smallholder's house.

With respect to changes to plot size after reallocation, from the 30 households taking part 12 households gained additional area and the sizes of

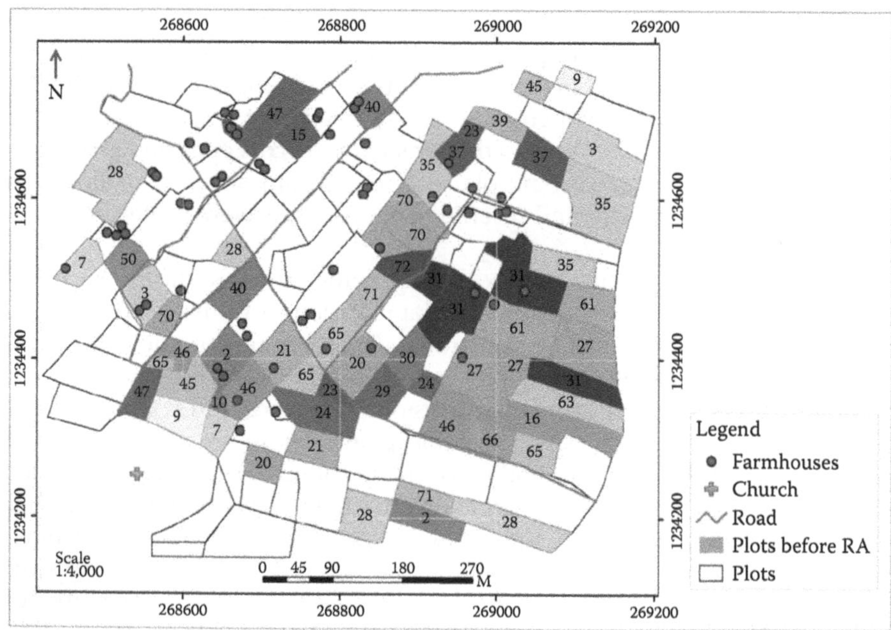

FIGURE 10.2
Farms, houses, and smallholder plot holdings prior to reallocation. RA, reallocation.

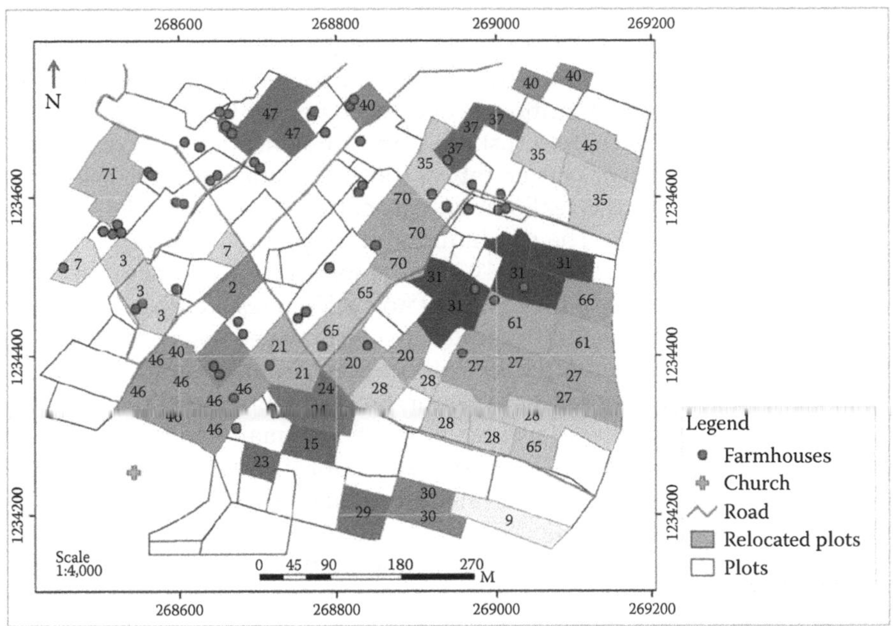

FIGURE 10.3
Farms, houses, and smallholder plot holdings after reallocation.

their aggregated plots increased. Meanwhile, 12 households lost area from the size of their previous plots. The remaining six households showed no change in their area holding. The maximum and minimum gain areas after reallocation were 0.083 and 0.003 ha, respectively. The maximum and minimum areas lost were 0.101 and 0.006 ha. The reason for this seemingly neutral result is suggested to be the absence of a land bank: the range of parcel combinations available was limited. At any rate, increases and decreases in parcel size are obviously only one criterion.

Regarding numbers of plots, for 19 out of 30 households the total number of plots decreased. Meanwhile, for 11 households there was no change. From those with a decrease in plot numbers, nine gained area, six lost areas, and four neither gained nor lost. From the 11 whose number of plots did not change, 3 gained area, 6 lost area, and 2 neither gained nor lost. Therefore, through another lens, from the 30 reallocated smallholders 13 gained area, 11 lost area, and 6 neither gained nor lost. Therefore, the reallocation process was considered advantageous for the majority of smallholders in terms of both area gained and decrease in total number of plots per smallholder (note: the newly allocated plots for single households, if now neighboring, were counted as a single plot).

Regarding plot concentration around farmhouses, this indicator was calculated by multiplying the area of reallocated plots by the distance from the farmhouses. From the total of 16 households who had farmhouses and reallocations, plots were concentrated for 14. Among these 14, 9 gained area, 4 lost area, and 1 neither gained nor lost area. The two smallholders whose plot concentration around the farmhouses decreased neither lost nor gained area. Overall, nine gained area, four lost area, and three neither gained nor lost. Therefore, regarding plot concentration the reallocation process was considered advantageous: the reallocation concentrated designed plots for the majority of smallholders. The maximum and minimum distance decreases after reallocation were 522.6 and 41.25 m, respectively. The maximum and minimum distance increases were 221.19 and 8.97 m, respectively.

Regarding plot shape, this indicator was calculated using the area to perimeter ratio before and after the reallocation. An increase in the area to perimeter ratio is considered positive for production: it generally means a more efficient plot shape for farming. From the 30 smallholders, the area to perimeter ratio increased for 15 and decreased for the remaining 15. For the 15 smallholders whose area to perimeter ratio increased, only 6 smallholders gained area from the reallocation, 7 smallholders lost area, and 2 smallholders neither gained nor lost. Meanwhile, from the total of 15 smallholders whose area to perimeter ratio decreased 6 smallholders gained area, 5 smallholders lost area, and 4 smallholders neither gained nor lost. Overall, the plot shape indicator produced a neutral result: the result was highly dependent on the farm areas after the reallocation.

Regarding the indicator dealing with closeness of the reallocated plots to the farmhouses, distance was decreased for 12 smallholders out of the 16 who had reallocations. Among these 12 smallholders, 8 gained area, 3 lost area, and 1 neither gained nor lost. Of the four smallholders whose distance between plots and farmhouses increased, one gained area, one lost, and two smallholders neither gained nor lost. Overall, nine smallholders gained area, four lost, and three neither gained nor lost. Therefore, the process was considered advantageous in terms of the closeness of designed plots to smallholders' houses: the reallocation shortened the distance traveled by smallholders.

Discussion

The results and their implications are discussed under the following themes: (1) fragmentation in Ethiopia; (2) contextual alignment between the Netherlands and Ethiopia; (3) outcomes from the showcase procedure and results; and (4) overall opportunities, challenges, and demands for application of land consolidation for Ethiopia. These themes relate to the overarching aim of this chapter and reveal key issues emerging throughout the study.

Regarding plot fragmentation in Ethiopia, against international comparisons (FAO 2003), the collected data and results reveal small plots sizes and high levels of parcel disaggregation for smallholders in the case area. Smallholders on average had 4.75 plots in different locations with an average size of 1.22 ha. In addition, many farm plots had farmhouses: many had built their own houses close to those of parents, brothers, and sisters. This is a preferred part of the livelihood and also a product of inheritance processes: most of the households settle in one place with their extended families surrounding them. Plots are distributed by gift or inheritance. Over time, plots are further subdivided to support new family offshoots. This is considered the main reason for fragmentation in the specific study area, and Ethiopia more generally.

Regarding the alignment of contextual conditions between Ethiopia and the Netherlands, obviously there are considerable political, economic, social, and environmental differences, not the least being that fragmentation was much higher in the study area than was ever experienced in the Netherlands. In addition, the smallholder interviewees, while concerned with production levels, appeared quite averse to the land consolidation concept. It needs to be assessed further whether the adversity is more widespread and in any way comparable to any community concerns experienced in the Netherlands. Meanwhile, there is alignment with the current agenda of the Ethiopian central government and the early land consolidation motivations of the Dutch: a focus on improving the income status and working conditions of rural dwellers is evident in both cases. The Dutch approach also sought to improve infrastructure and agricultural production. Ethiopia has these same objectives; however, the use of land consolidation to support these activities is not yet apparent. Arguably, this is because all land is state owned: acquiring land for infrastructure or large-scale land investments requires less prior and informed consent and participation from existing land users. At any rate, all European countries undertaking land consolidation activities have different underlying objectives for doing so. What is perhaps more important is that the procedures used for implementation were more or less similar. In this light, it seemed reasonable to undertake the showcasing exercise.

Regarding the showcase, the *requirements gathering* process was seen to be reasonably efficient: the small research group was able to acquire data on 70 plots in a participatory fashion over a short period. It is felt that the procedure could be easily repeated and scaled and does not require high levels of technological capacity. In contrast, the *design procedure* was based on approaches developed in the Netherlands, and advice from the Netherlands Kadaster: the algorithms and technology, including ArcGIS, were central. The reallocation process was executed based on the premise that smallholders had the right to exchange plots; however, it had no actual effect on the land use rights of smallholders. Overall, most reallocated smallholders got comparable and improved plot structures: a land bank would have helped those where this was not the case. The approach required reasonably sophisticated technical understandings of land consolidation, algorithms, and GIS. To be undertaken at scale would require significant amounts of training and capacity development within the Ethiopian land management sector. This appears to go against the procedures used in Ethiopia's rural land certification projects: lower skilled local land committees were effectively used to deliver millions of titles. Therefore, in line with the ideas of Thapa and Niroula (2008), simplified land consolidation processes need to be sought. The *testing and refinement* procedure was more limited: resource limitations meant that the reallocation design could not be taken back to the study location and approved, rejected, or further refined with the participants. At any rate, the design was realized in a virtual sense and could be evaluated against the predefined criteria. The resultant reallocation saw many smallholders improving their plot concentration and plot sizes and reducing the distance between their farmhouses and designed plots.

Regarding *opportunities for further development,* the process is considered innovative and potentially aligns with fit-for-purpose land administration ideals (FIG 2014): the method considers important local landscape elements such as vegetation along fence lines, and the fact that smallholders might have multiple farmhouses on separate plots. Importantly, the approach is also repeatable and can potentially be scaled. More importantly, there was real improvement to farm size and structure resulting from the reallocation: it could be expected that the reallocation, if actually applied, would have a positive impact on agricultural productivity. If the Dutch experience is to be considered, the approach could also result in faster agricultural restructuring, adoption of new agricultural technologies, and improved productivity and income. However, such positive sentiments should be tempered, as a number of concerns remain.

Regarding overall *challenges to further development,* the approach demands that some smallholders lose area on reallocation. Given the results from smallholder interviews, and local knowledge regarding relationships with land, this would most likely be unacceptable. In a usual setting, a land bank would be used to offset any losses in total area; however, no land bank was incorporated into this specific design and the results of the showcase were thus impacted. A potential solution would be to extend Ethiopia's existing land banks and utilize them in land consolidation activities: the Dutch experience suggests that successful implementation is dependent on land banking. As already noted, Ethiopian land banks are not multipurpose and are solely used to support large-scale land investment for foreign and domestic interests. Additionally, although the approach utilized substantial levels of technology, it did not utilize the reallocation programs and algorithms already available in ArcGIS: the approach could potentially be made simpler. Meanwhile, taking the Dutch experience into consideration, when applied, the approach could also result in a decrease in biodiversity and natural landscapes, a drawn out implementation period and incur high overheads.

Despite these challenges, land consolidation as an option for Ethiopia demands further research attention, if only because other food security interventions are not delivering enough improvements. For smallholders, fertilizer programs have been trialed; however, the relevance for staple crops like teff is highly questionable (Kraaijvanger and Veldkamp 2014). In addition, enforced production of specific crops has been another measure; however, this undermines access to specific nutrients for populations in those areas. Another strategy is to use more land; however, in highland Ethiopia a shortage already exists. Resettlement to lowlands is another possibility; however, it is highly disruptive and deeply unpopular (Ambaye 2013). Technology investment is encouraged, but it is ultimately impeded: concerns still remain about the security of the rural land tenure system. Through the work detailed in this chapter, it is shown that, at least from a technical perspective, land consolidation can be applied in Ethiopia. However, it needs to be stressed that potential social and environmental impacts of the activity, and scalability issues, need further thorough investigation. It is likely that a far simpler and less technology-intensive land consolidation tool would be needed, one aligned with the low-cost land certification program used successfully in many parts of rural Ethiopia and the fit-for-purpose land administration more generally.

Conclusions

The chapter aimed to investigate if and how conventional land consolidation procedures could be reengineered to fit the Ethiopian context. To this end, a case area in Ethiopia's Amhara region was selected, field data were collected from local community and government sources, a reallocation design procedure was adapted from the Netherlands context, the procedure was applied, and the subsequent redesign was evaluated against several present criteria.

The study confirms that land fragmentation is prevalent in the study area and impedes agricultural development. However, land consolidation is not a straightforward solution: a range of social, legal, and technological constraints are apparent. Families prefer to be clustered closely on small farms, and disparate plots are seen as a pertinent risk management strategy. More fundamentally, the current land tenure system does not enable plot exchange, as was fabricated in the showcase. The lack of a multipurpose land bank and minimal local land consolidation technical capacity present other challenges to the approach. For these reasons, fertilizer programs, enforced land uses, and resettlement activities are the focus of government and donor food security interventions: land consolidation would require a long-term vision, greatly enhanced local technical capacity or far simpler tools, an unwavering implementation phase, and patience.

At any rate, from a mere technical perspective the study shows that conventional land consolidation tools can be applied in the Ethiopian context as a parcel-based land tenure system exists. Moreover, from the reallocation process, and even without a land bank, the majority of smallholders, with more than one plot, received larger plot sizes, improved plot shapes, higher plot concentration, and improved plot proximity to their houses. In this respect, it is recommended that the results of the study are disseminated to the smallholders and government agencies: they could form the basis of further piloting work that potentially incorporates infrastructure provision and takes into account social, economic, and environmental aspects. However, the approach offered here is technologically intensive and requires large amounts of highly trained staff to execute at scale. In this respect, there is impetus to explore the creation of simpler tools for smallholder land consolidation, ones that are more amenable to participation and ownership by local communities and more aligned to fit-for-purpose land administration ideals.

References

Alemu, G. 2012. *Rural Land Policy: Rural Transformation and Recent Trends in Large-Scale Rural Land Acquisitions, Ethiopia European Report Development* (p. 28). Brussels, Belgium: Overseas Development Institute (ODI) in partnership with the Deutsches Institut für Entwicklungspolitik (DIE) and the European Centre for Development Policy Management (ECDPM).

Ambaye, D. 2013. *Land Rights and Expropriation in Ethiopia*. Doctoral diss. Stockholm, Sweden: KTH.

Ayranci, Y. 2007. Re-allocation aspects in land consolidation: a new model and its application. *Journal of Agronomy*, 6(2), 270.

Belay, A. 2010. *The Effects of Rural Land Certification in Securing Land Rights: A Case of Amhara Region, Ethiopia*. Masters thesis. Enschede, The Netherlands: International Institute for Geo-Information Science and Earth Observation.

Bennett, R.M., and Alemie, B.K. 2015. Fit-for-purpose land administration: lessons from urban and rural Ethiopia, Survey Review (In Press). http://dx.doi.org/10.1179/1752270614Y.0000000149.

Bergh, v.d.S., 2004. *The History of Land Consolidation in Netherlands*. PhD thesis. Wageningen, The Netherlands: Wageningen University.

Bonfanti, P., Fregonese, A., and Sigura, M. 1997. Landscape analysis in areas affected by land consolidation. *Landscape and Urban Planning*, 37(1), 91–98.

Buis, A., and Vingerhoeds, R. 1996. Knowledge-based systems in the design of a new parcelling. *Knowledge-Based Systems*, 9(5), 307–314.

Cay, T., Ayten, T., and Iscan, F. 2010. Effects of different land reallocation models on the success of land consolidation projects: social and economic approaches. *Land Use Policy*, 27(2), 262–269.

Crecente, R., Alvarez, C., and Fra, U. 2002. Economic, social and environmental impact of land consolidation in Galicia. *Land Use Policy*, 19(2), 135–147.

CSA. 2012. *The Inter-Censal Population Survey (ICPS) Based on Statistical Report of the 2007 Population and Housing Census*. Federal Democratic Republic of Ethiopia Population Census Commission. Addis Ababa, Ethiopia: CSA.

Damen, J. 2004. *Land Banking in the Netherlands in the Context of Land Consolidation.* International Workshop: Land Banking/Land Funds As an Instrument for Improved Land Management for CEEC and CIS. Tonder, Denmark.

Deininger, K.W., and Byerlee, D. 2011. *Rising Global Interest in Farmland: Can It Yield Sustainable and Equitable Benefits?* Washington DC, United States: World Bank Publications.

Demetriou, D., Stillwell, J., and See, L. 2011. LandSpaCES: a spatial expert system for land consolidation. In: S. Geertman, W. Reinhardt, and F. Toppen (Eds.), *Advancing Geoinformation Science for Changing World* (pp. 249–274). Dordrecht, The Netherlands; Berlin, Germany; Heidelberg, Germany: Springer.

Demetriou, D., Stillwell, J., and See, L. 2012. Land consolidation in Cyprus: why is an integrated planning and decision support system required? *Land Use Policy*, 29(1), 131–142.

De Vos, W. 1982. Allocation in land consolidation projects in the Netherlands with the aid of an automated system. *Surveying and Mapping*, 42(4), 339–345.

Essadiki, M. 2002. New method for land reallocation by using a geographic information system. In: *Proceeding of XXII FIG International Congress*, Washington, DC.

Essadiki, M., and Ettarid, M. 2002. *Optimization of Technical Steps of Rural Land Consolidation Project using a GIS.* International Symposium on GIS, Istanbul, Turkey.

Essadiki, M., Ettarid, M., and Robert, P. 2003. *Optimisation of Technical Steps of a Rural Land Consolidation Using a Geographic Information System: Land Reallocation Step.* FIG Working Week 2003, Paris, France.

FAO. 2003. The design of land consolidation pilot projects in Central and Eastern Europe. In: *FAO Land Tenures Studies FAO Land Tenure Studies* (Vol. 6). Food and Agriculture Organization of the United Nations.

FAO. 2013. *National Resource Management for Food Security in the Context of the Post 2015 Development Agenda: Empowering Small Scale Food Producers and Food Insecure Communities to Be Agents of Change.* Committee on World Food Security: Special Event Concept Note, Rome, Italy.

FDRE. 2003. *Rural Development Policy and Strategies.* Addis Ababa, Ethiopia: Federal Democratic Republic of Ethiopia.

FDRE. 2010. *Growth and Transformation Plan 2010–2014.* Addis Ababa, Ethiopia.

FDRE. 2013. *Amhara National Regional State*, accessed November 9, 2013, http://www.ethiopia.gov.et/web/pages/StateAmhara.

FIG. 2014. *Fit-for-Purpose Land Administration*, FIG Publication 60. Copenhagen, Denmark.

Gebreselassie, S. 2006. *Land, Land Policy and Smallholder Agriculture in Ethiopia: Options and Scenarios.* Future Agricultures, Discussion Paper 8, Brighton, United Kingdom.

Hartvigsen, M. 2005. *Land Consolidation Pilot Projects in Eastern Europe.* Budapest, Hungary: Land Consolidation Conference.

Hoben, A. 2000. *Ethiopian Rural Land Tenure Policy Revisited: A Symposium for Reviewing Ethiopia's Socioeconomic Performance 1991–1999*, Paper number 22. InterAfrica Group. April 26–29, 2000, Addis Ababa, Ethiopia.

Hupfeld, W. 1971. Ein Beispiel zur mathematischen Planungsrechnung. *Zeitschrift für Vermessungswesen*, 2, 61–65.

Jansen, L.J.M., Karatas, M., Küsek, G., Lemmen, C., and Wouters, R. 2010. The computerised land re-allotment process in Turkey and the Netherlands in multipurpose land consolidation projects. 24th FIG Congress. April 11–16, Sydney, Australia.

Jansen, L.J.M. and Wubbe, M. 2012. Better agricultural conditions by improving land management (G2G/11/RM/8/1), ANCPI, Agentschap NL EVD International, Internal Report. Netherlands Embassy.

Kadaster. 2012. *Herverkaveling, WILG projecten en Verkavelen voor groei*, accessed October 20, 2013, http://www.kadaster.nl/web/Jaarverslag-2012/Inwinning-en-registratie/Herverkaveling.htm.

Kik, R. 1971. Een methode voor het vervaardigen van een voorlopig toedelingsplan voor een ruilverkaveling (A method to construct a preliminary re-allotment plan). *Nederlands Geodetisch Tijdschrift*, 207–215.

Kik, R. 1979. *Toedelingsonderzoek in de voorbereidingsfase van een ruilverkaveling. Cultuurtechnisch Tijdschrift*, 1979(4).

Kool, S. 2013. *Voluntary Re-Allotment in the Netherlands: A Case Study about the Processes of Voluntary Re-Allotment in Heusden, Epe-Vaassen and Kempen-Broek*. Masters thesis. Wageningen, The Netherlands: Wageningen University.

Kraaijvanger, R., and Veldkamp, T. 2014. Grain productivity, fertilizer response and nutrient balance of farming systems in Tigray, Ethiopia: a multi-perspective view in relation to soil fertility degradation, *Land Degradation and Development*, DOI: 10.1002/ldr.2330.

Lemmen, C., Jansen, L., and Rosman, F. 2012. *Informational and Computational Approaches to Land Consolidation*. FIG Working Week May 2012, Rome, Italy.

Lemmen, C., and Sonnenberg, J. 1986. A model for allocation and adjustment of lots in land consolidation: new development in the Netherlands. Federation Internationale des Geometres, XVIII International Congress, June 1–11, Toronto, Canada.

Louwsma, M. 2013. *Participation in re-allotment*. Kadaster, Apeldoorn, The Netherlands.

LTO Nederland. 2013. *Verkavelen voor groei*. Kadaster, Apeldoorn, accessed May 13, 2015, http://www.verkavelenvoorgroei.nl/wp-content/uploads/2013/09/130919-Verkavelen-voor-groei.pdf.

LTO Nederland. 2014. *Proces vrijwillige*, accessed November 10, 2013, http://www.verkavelenvoorgroei.nl/hoe-gaat-verkavelen.

Mihara, M. 1996. Effects of agricultural land consolidation on erosion processes in semi-mountainous paddy fields of Japan. *Journal of Agricultural Engineering Research*, 64(3), 237–247.

Nega, B., Adenew, B., and Sellasie, S. 2003. *Current Land Policy Issues in Ethiopia Land Reform Special Edition* (Vol. 3, pp. 103–154). Addis Ababa, Ethiopia: Ethiopian Economic Policy Research Institute.

Niroula, G.S., and Thapa, G.B. 2005. Impacts and causes of land fragmentation, and lessons learned from land consolidation in South Asia. *Land Use Policy*, 22(4), 358–372.

Pašakarnis, G., and Maliene, V. 2010. Towards sustainable rural development in Central and Eastern Europe: applying land consolidation. *Land Use Policy*, 27(2), 545–549.

Riddell, J., and Rembold, F. 2002. *Farm Land Rationalization and Land Consolidation: Strategies for Multifunctional Use of Rural Space in Eastern and Central Europe*. International Symposium on Land Fragmentation and Land Consolidation in CEEC: A Gate Towards Sustainable Rural Development in the New Millennium, Munich, Germany.

Rienks, W., Meulenkamp, W., Olde Loohuis, R., and van Rooij, B. 2009. *Agricultural Atlas of the Netherlands, the Dutch Agricultural Sector on the Map*. ROM3D, Hengevelde, The Netherlands.

Rosman, F. 2012. *Automated Parcel Boundary Design Systems in Land Consolidation.* FIG Working Week May 2012, Rome, Italy.

Sabates-Wheeler, R. 2002. Consolidation initiatives after land reform: responses to multiple dimensions of land fragmentation in Eastern European agriculture. *Journal of International Development*, 14(7), 1005–1018.

Schrader, B. 1971. Ablaufplanung und mathematische Optimierung bei der Flurbereinigung. *Vermessungstechnische Rundschau*, 33.

Sklenicka, P. 2006. Applying evaluation criteria for the land consolidation effect to three contrasting study areas in the Czech Republic. *Land Use Policy*, 23(4), 502–510.

Teka, K., Van Rompaey, A., and Poesen, J. 2013. Assessing the role of policies on land use change and agricultural development since 1960s in northern Ethiopia. *Land Use Policy*, 30(1), 944–951.

Tenkanen, J. 1987. Computer-aided allocation of plots in land consolidation. *Journal of Surveying Science Finland*, 1987(2). 10–25.

Teshome, A., Graaff, J., Ritsema, C., and Kassie, M. 2014. Farmers' perceptions about the influence of land quality, land fragmentation and tenure systems on sustainable land management in the north western Ethiopian highlands. *Land Degradation and Development*, DOI: 10.1002/ldr.2298.

Thapa, G.B., and Niroula, G.S. 2008. Alternative options of land consolidation in the mountains of Nepal: an analysis based on stakeholders' opinions. *Land Use Policy*, 25(3), 338–350.

Thomas, J. 2004. *Modern Land Consolidation—Recent Trends on Land Consolidation in Germany*. Symposium on Modern Land Consolidation, FIG Commission 7. September 10–11, 2004. Volvic (Clermond-Ferrand), France. FIG, Copenhagen, Denmark.

Thomas, J. 2006a. Property rights, land fragmentation and the emerging structure of agriculture in Central and Eastern European countries. *Electronic Journal of Agricultural and Development Economics Food and Agriculture Organisation*, 3(2), 225–275.

Thomas, J. 2006b. Attempt on systematization of land consolidation approaches in Europe. *Zeitschrift für Vermessungswesen*, 3, 156–161.

Touriño, J., Parapar, J., Doallo, R., Boullón, M., Rivera, F.F., Bruguera, J.D. et al. 2003. A GIS-embedded system to support land consolidation plans in Galicia. *International Journal of Geographical Information Science*, 17(4), 377–396.

Van den Brink, A. 2004. *Land Consolidation and the Emergence of the Metropolitan Landscape*. Symposium on Modern Land Consolidation, FIG Commission 7. September 10–11, 2004. Volvic (Clermond-Ferrand), France. FIG, Copenhagen, Denmark.

Van den Noort, P.C. 1987. Land consolidation in the Netherlands. *Land Use Policy*, 4(1), 11–13.

Van Dijk, T. 2003. Scenarios of Central European land fragmentation. *Land use policy*, 20(2), 149–158.

Vitikainen, A. 2004. An overview of land consolidation in Europe. *Nordic Journal of Surveying and Real Estate Research*, 1(1): 26–44.

WFP. 2014a. *Global Food Security Update: Tracking Food Security Trends in Vulnerable Countries* (Issue 13), February 2014. Rome, Italy: World Food Program.

WFP. 2014b. *Hunger Map 2014*. Rome, Italy: World Food Program.

Wu, Z., Liu, M., and Davis, J. 2005. Land consolidation and productivity in Chinese household crop production. *China Economic Review*, 16(1), 28–49.

Yimer, F. 2014. *Fit-for-Purpose Land Consolidation: An Innovative Tool for Re-Allotment in Rural Ethiopia*. MSc thesis. Enschede, Netherlands: University of Twente.
Zhengfeng, Z., and Wei, Z. 2007. Effects of land consolidation on ecological environment. *Transactions of the Chinese Society of Agricultural Engineering*, 2007(8).

11

Dynamic Nomadic Cadastres

Monica Lengoiboni, Arnold Bregt, and Paul van der Molen

CONTENTS

Introduction

A cadastre is a parcel-based and up-to-date land information system containing a record of interests (viz., rights, restrictions, and responsibilities) together with a geometric description of land parcels and their boundaries (FIG 1995). A cadastre is seen as the database establishing the core of a broader land administration system, aimed at determining, recording, and informing about the ownership, value, and use of land when implementing land management policies (UN 1996). Rights to land might be held under various legal regimes, such as statutory law, common law, and customary law. The cadastral process includes adjudication of existing land right, however, mainly when regulated in a formal law (e.g., "ownership" and "freehold"). In areas where customary tenure prevails, cadastres hardly have coverage because the definitions of land rights, allocation, and protection are outside the formal law. Adjudication also assumes the existence of land parcels and boundaries. This causes a problem where either boundaries are conceptually not known in local cultures or land rights or land uses show a dynamic aspect, for example, if they change from time to time. This is the situation that pastoralists face: their land tenure developed within local customary traditions is often not recognized in statutory law. The dynamic nature of their land use does not comply very well with the adjudication requirements of determinable owners, land parcels, and boundaries.

Pastoralists' rights to land, based on their customary tradition, require extensive livestock mobility between seasonal resources for effective productivity. Established patterns of seasonal mobility and migration routes link dry and wet season grazing areas (Blench 2001; Lengoiboni, Bregt, and van der Molen 2010; Liao et al. 2014), making it possible to predict to a certain degree where and when pastoralists are likely to be. Moreover, customary rights of access to dry season grazing areas are traditionally based on reciprocal arrangements on the use of property rights, for example, between agriculturalists and pastoralists (McCarthy et al. 1999). Such flexible tenure arrangements allow pastoralists to exploit the available resources across various agroecological conditions, thereby reducing their levels of vulnerability (Niamir-Fuller 2005; Nori 2007). This system of pastoralists' seasonal mobility and interaction between pastoralists' and farmers' tenures and land uses has for a long time not been understood by policy makers and practitioners responsible for interpreting and implementing rights to land. Policies and strategies introduced in pastoral areas have disregarded the importance of livestock mobility, while perception of pastoralism as a creator of livelihoods, a driver of economic growth, and a component of social identity has been ignored.

Under a statutory/formal land tenure system, property rights commonly occur in the forms of freehold, leasehold, easements, and usufruct/profit à prendre, among others. The types of rights within these tenures vary from full ownership to limited rights, for example, of access and use. Although formal rights apply to private parcels, pastoralists' customary rights differ especially in terms of fluidity of where and when their rights apply in a geographical space and time, that is, seasonal access to migration corridors and to grazing areas. Sedentarization policies led to new forms of tenures being designed to position pastoralists' tenures to adjust to the requirements of formal tenure (Coldham 1979). A variety of tenures were implemented in various countries to support pastoral systems, for example, grazing reserves in Nigeria; ranch-type tenures in Tanzania, Burkina Faso, Senegal, and Kenya; and the tribal grazing lands under the Tribal

Grazing Lands Policy (TGLP) in Botswana, among others (Awogbade 1987; Rietbergen-McCracken and Abaza 2014). In Kenya, the group ranch tenure was created by the Kenyan government in the 1960s. A group ranch is a private statutory tenure in which the members jointly own a freehold title to large tracts of land, which they use communally. These group ranches were systematically created, one next to another in some pastoral areas. Policy makers viewed group ranches as a temporary measure to accommodate pastoralists' unique land tenure and uses in the formal system, with the expectation that right holders would soon recognize the advantages of and transform into individual land titles (Okoth-Ogendo and Oluoch-Kosura 1995)

For pastoralists, spatial fixity introduced by the group ranch concept in Kenya brought about undesired outcomes. Environmental degradation due to herd pressure and overgrazing combined with inefficient management eventually led to the disintegration and subdivision of many group ranches into individual parcels (Galaty 1988; Kimani and Pickard 1998; Mwangi 2007; Rutten 1992; Western, Groom, and Worden 2009). Meanwhile, the resilience of pastoralism and maintained seasonal migrations amid the challenges of land fragmentation, population increase, and reduced land for grazing and a growing understanding of the importance of mobility in pastoral systems triggered a paradigm shift toward its recognition and a movement toward supporting their migratory land use system in the National Land Policy (Kenya 2009). The Constitution of Kenya (2010a) recognizes customary land rights under community land. Community land is defined as areas where no adjudication and demarcation of individual or group tenures have occurred (Kenya 2013). The Community Land Bill (Kenya 2013) lays down the procedures for adjudication, demarcation, and delineation of boundaries on community land. This progression toward formal recognition of customary and pastoralists' land rights in the statutory framework may just be a first step. What is also needed is an investigation into the forms of tenure that correspond to pastoral dynamic rights within the statutory tenure system.

This research builds on earlier papers that showed that pastoralism is still actively practiced and how affected parties solve problems that arise when pastoralists' seasonal land use under customary law overlaps with private tenures under statutory law (Lengoiboni, Bregt, and van der Molen 2010; Lengoiboni, van der Molen, and Bregt 2011). Both papers give evidence that regulation, documentation, and protection of pastoralists' dynamic land rights are urgent. In this chapter, we seek to find ways on how pastoralist tenures could be accommodated in the formal law such that they can also be eligible for registration in a cadastre—and, therefore, a "dynamic nomadic cadastre." In this way, adjudication and registration would provide security of tenure not only to sedentary citizens but also to nomadic citizens.

Theoretical Background

Formalization of land rights often failed to deliver expected results in Africa because they could not accommodate the customary tenure system and the way people think about and use land. Formalization of property rights creates exclusive forms of ownership of resources and ignores overlapping interests (Meinzen-Dick and Mwangi 2009), which is crucial for pastoralists. Pastoralists' practice of repeatedly renegotiating temporary and flexible access rights to resources is challenged as land is increasingly being surveyed, demarcated, and allocated (Homewood et al. 2004). Pastoralists' social and economic welfare has declined due to external pressures as they are unable to respond appropriately to meet the requirements of their traditional mode of production (FAO 1999; Swallow and McCarthy 1999). Pastoralists' land rights on migration corridors and grazing areas often remain undocumented and therefore off the cadastres.

Cadastres, or land administration systems in the narrow sense, are just databases: they do not define ownership or boundaries in themselves. Their aim is to archive which tenures exist on the ground, after a process of adjudication. The institutional context is paramount in providing legal and societal meaning to this information. Thus, dynamic nomadic cadastres are meaningless when "dynamic nomadic land rights" are not defined and regulated in the formal environment. Of course, they might exist in the nonformal environment—this issue is not at stake. However, to be eligible for registration and acquire authoritative protection, the institutions (e.g., law and governance) need to recognize their existence. This is a major area of attention in this chapter.

Western African countries such as Burkina Faso, Mauritania, Mali, Guinea, and Niger have adopted certain legal provisions to protect pastoralists' land rights. A common feature in their pastoral codes is that pastoralists' access rights to migration corridors are conferred based on an open access regime. Herders also have the right to move animals at local and regional scale within their national territories, as well as across national borders. In general, the form of tenure is public domain for state or local government, and/or communal use. Rights of access to grazing areas are not mentioned in the laws; so far, they remain unregulated. In any case, the inclusion of pastoralists' land rights in a statutory system appears to be possible. The West Africa's approach to securing pastoralists' land rights guides our theoretical background to finding how pastoralists' dynamic tenure could be accommodated in the formal law.

In Kenya, the process of recognizing pastoralists' land rights in the formal system has begun. The National Land Policy (2007) explicitly states that "the government shall institute alternative methods of registration that define individual rights in pastoral communities while allowing them to maintain their unique land use systems and livelihoods." This makes it extremely interesting to investigate how pastoralists' land rights could be accommodated. An examination of the Kenyan statutory framework leads us to conclude that the following tenure options are available:

Freehold: An ownership of land rights in perpetuity. It is a statutory right described in Kenya's Registered Land Act (Cap 300). Individuals have rights of ownership and have responsibilities and restrictions placed by the state or other third parties (Dale and McLaughlin 2000).

Lease: Ownership of rights for a limited period. These are recognized by the Registered Land Act.

Easement: "a right attached to a parcel of land which allows the proprietor of the parcel either to use the land of another in a particular manner or to restrict its use to a particular extent, but does not include a profit" (Kenya 1963). It is a limited land right, a nonownership right recognized by the Registered Land Act. These limited land rights can coexist with other interests in land.

Profit (profit à prendre): The "right to go on the land of another and take a particular substance from that land, whether the soil or products of the soil" (Kenya 1963). It is a limited land right recognized by the Registered Land Act and can also coexist with other interests in land.

Open access: This mode of access to land exists where there is no defined group of owners. Benefits are available to anyone and there are no duties or obligations (Dale and McLaughlin 2000).

Negotiations with landowners: This mode of access to land is derived from current practice in the study area, whereby pastoralists negotiate for access with private landowners. Entering private land without the permission of the landowner is considered to be trespassing and is liable to punishment by law (Kenya 1963).

Customary land rights: Customary land rights are accommodated in Kenya's Trust Lands Act (Cap 288). Section 69 of Cap 288 allows the occupiers of trust lands to enjoy land rights according to their customary law, including any subsequent modifications of the land rights, but only as long as such rights do not conflict with any of the provisions of the act or rules made under it, or to the provisions of any other law currently in force.

Reserved land (government land): Land that the government reserves for public benefit. State agencies set rules for access and use of the reserved land, and individuals have duties to respect those rules. Reserved land is recognized by the Land Acquisition Act (Cap 295). This act provides for the compulsory acquisition of land by the government to be reserved for public benefit, such as land for schools, hospitals, parks, and so on. This mode of access to land was suggested by land professionals during the early stages of testing the questionnaire. Reserved land was therefore included as an additional tenure option in the decision matrix.

This theoretical background leads us to the question "which of the above land tenure options can best support the dynamic nature of pastoralists' seasonal land rights in a dynamic cadastre?" From a system perspective, the dynamic nomadic cadastre may benefit from recent developments in domain modeling. The Social Tenure Domain Model (Lemmen 2012) is a land administration tool that enables the capture of all land rights as they exist in reality, including all forms of rights to land, all kinds of properties, and all kinds of spatial objects, regardless of their level of formality. By capturing an inventory of land rights as they exist, such as the spatial and temporal aspects of pastoralists' land rights revealed in this chapter, the model could provide a technical basis for the dynamic nomadic cadastre in the formal land administration system.

Methodology

The objective of this study is to find out what tenure options have the potential to support pastoralists' dynamic land rights in the statutory tenure system. Land professionals working in the field of land administration in Kenya were approached to map out and weigh potential tenure solutions grounded in Kenyan law. The land professionals were selected for this study for three main reasons: first, because land administration is almost always restricted to statutory land tenures (van der Molen 2003); second, land tenure depends on registration of land rights, which is an instrument for implementing land policies (i.e., through adjudication and registration of land rights), and the role of the government is to lay down a legal framework for administering the statutory land rights (van der Molen 2003); and, third, in the cadastral processes of adjudication, survey, and registration of statutory land rights land professionals are directly involved in introducing Western-style ownership tenures to replace customary tenures.

Land professionals from the Ministry of Lands (MoL) and district local governments—the county councils (CCs) in Kenya—were approached to participate in this study. The MoL is responsible for land policy, physical planning, land rights adjudication and settlement, undertaking land surveys and mapping, registration of land rights, land valuation, and administration of state and community land. The MoL consists of four main departments: Lands, Physical Planning, Survey, and Land Adjudication and Settlement.

Directors of these departments at the national level (Nairobi), technical staff at provincial level (Nakuru–Rift Valley province), and staff at county level (Samburu, Isiolo, and Laikipia counties—districts at the time of the interview) were also interviewed.

CCs are mandated to alienate land from the community lands (formerly trust lands) within their jurisdiction and have their own land surveyors. The CC land surveyors were included as land professionals in this study because the alienation of sections of community lands, for example, for private purposes, may interfere with pastoralists' customary land rights. From CCs, surveyors, adjudicators, physical planners, and land officers were interviewed.

Further, the chairman of the Institution of Surveyors of Kenya (ISK) was also interviewed as one of ISK's objectives is to contribute to the development of international and national policies and legal frameworks, strategies, and plans in land management in a manner that facilitates sustainable development.

Fieldwork was carried out in July 2010. Appointments were made in advance to invite the land professionals to participate in the study. They evaluated the tenure options that could be used to secure (1) the migration corridors that pastoralists use for their seasonal movements and (2) dry season grazing areas, where pastoralists graze their livestock during the dry seasons. Of the 18 land professionals approached to complete the questionnaire, 13 completed the questionnaire, 2 were unavailable (out of office), and 3 declined to complete the questionnaire. The work experience of the land professionals varied from a few years to more than 20 years: 0–4 years: three respondents; 5–9 years: one respondent; 10–14 years: one respondent; 15–19 years: five respondents; and >20 years: three respondents.

Questionnaire

The questionnaire was administered in the form of a structured interview and contained two parts. The first part dealt with the tenure options that could support migration corridors and dry season grazing that are confined within the pastoral areas. We refer to pastoral areas as context 1.

The second part of the questionnaire dealt with whether pastoralists should be allowed to migrate with their livestock in search of dry season pastures in non-pastoral areas, that is, where pastoralists' tenures seasonally overlap with private tenures (farming areas and on government land, i.e., wildlife parks and forests) in northern Kenya. The respondents answered closed questions with "yes," "no," or "do not know" as their preferred answer. Those answering yes or do not know went ahead to evaluate the statutory tenure options that have potential to support pastoralists' seasonal land rights in the non-pastoralist areas. We refer to the non-pastoralist areas as context 2.

In this study, we evaluated various types of tenure options: statutory tenures, tenures based on customary rights, and negotiated agreements. Tenure options were evaluated because the extent to which people can access and use land is often built into the land tenure system. Eight tenure options were included in the evaluation, as follows:

1. Freehold
2. Leasehold
3. Easement
4. Profit
5. Open access
6. Negotiations with landowners
7. Customary land rights
8. Reserved land (government land)

These tenure options were evaluated for contexts 1 and 2. In some situations, tenure options were omitted as follows:

Profit: This was omitted for migration corridors because pastoralists require access to travel across them.

Easement: This was omitted for grazing areas because pastoralists require access to remain on them for a period to reap benefits from the land.

Negotiations and reserved land: These were omitted for context 1 (pastoralist areas) because customary social systems continue to enable herders to maintain connectivity between seasonal resources where those resources are located within the pastoral areas. In other words, customary land rights and practices have not yet been extinguished, especially in community lands.

Open access: This was omitted for context 2 (non-pastoralist areas) with the assumption that most of the land is under some form of control by either individuals or the government.

Evaluating Tenure Options Using a Decision Matrix

The Pugh Decision Matrix was used in this research. It was developed by Stuart Pugh to help decision makers choose from a number of options (Pugh 1996). Multiple options are evaluated against each other using predetermined criteria to help decision makers select a satisfying or most promising option that is likely to result in successfully solving a problem. The Pugh Decision Matrix was also preferred because it does not require the criteria to be weighted. This means that all criteria are considered to be equally important and therefore have the same weight. The Pugh Decision Matrix requires establishing a baseline or a reference point, which can be one of the options (Pugh 1996). According to Pugh (1996), the baseline should be considered a somewhat average/neutral idea—neither the best nor the worst. In this study, customary tenure was used as the baseline, because it is the mode that pastoralists use to exercise their seasonal migrations and access dry season resources. Each option is rated on how well it meets each criterion in comparison with the baseline (Pugh 1996). Thus, if the weight assigned to all criteria is zero [0], [+1] score is assigned to options that would satisfy the criterion better than the baseline, a zero [0] is assigned to options that would satisfy the criterion to the same degree as the baseline, and [–1] is assigned to options that would not satisfy the criterion as well as the baseline. The option with the highest score is considered the most desirable option. However, according to Pugh (1996) the highest score might not guarantee the optimal solution but might give rise to a discussion, for example, about further development or modification of the option.

Criteria for Evaluating Appropriate Tenure Options

The criteria were listed in the left-hand column in the decision matrix. The criteria can be viewed as tools for assessing issues and considerations likely to affect an option's implementation and its feasibility in achieving the intended goals (Bardach 2009). The criteria are therefore a list of factors considered important when making a decision. In this study, the goals that the land tenure options should achieve are social, economic, legal, and other requirements related to the pastoralists' use of migration corridors and access to dry season grazing areas. Six basic criteria were used in this study. The criteria are as follows:

1. Administrative feasibility—to judge the ease of implementation of options
2. Economic benefits—to judge whether the option is likely to benefit the actors involved and has the lowest cost of implementation
3. Effectiveness—to judge whether the option is likely to contribute to security of access to migration corridors and dry season grazing areas and at the same time provide security of tenure to the actors involved
4. Equity—to judge whether the option is likely to take fairness into account regarding access to land (e.g., will some actors be left out in a given tenure option?)
5. Technical feasibility—to judge whether the resources and skills needed to implement the option are readily available
6. Legal acceptability—to judge whether the use of the option to support migration and dry season grazing areas is permissible under existing land laws

These criteria were adapted from the works of Bardach (2009) and Patton and Sawicki (1986), who give basic frameworks for the criteria used in policy analysis. The policy analysis criteria are considered relevant for use in this study because they relate to the factors considered when implementing the policies. As land tenure—through adjudication and registration of land rights—is a means by which land policies are implemented, the policy analysis criteria provide guidance for assessing the goals that the options should meet. It is therefore reasonable to assume that these criteria are directly relevant to the analysis of tenure options likely to support pastoralists' migratory land use. The respondents ranked the criteria against each tenure option according to their own judgments, while referring to migration corridors in the study area—derived by Lengoiboni et al. (2010).

Results

Tenure Options for Securing Pastoralists' Seasonal Land Rights

The judgments or opinions of the 13 land professionals are presented in this section. Their intersubjective opinions on the potential of various statutory tenure options to support pastoralists' seasonal land rights on migration corridors and in dry season grazing areas in contexts 1 and 2 are given. Varying proportions of land professionals provided their opinions on contexts 1 and 2.

Context 1: Dry Season Grazing within Pastoral Areas

All the 13 land professionals completed the questions in context 1. Respondents overwhelmingly favored nonownership forms of tenure to support pastoralists' land rights within the pastoralists' areas. Easements received the highest scores for migration corridors, whereas profit scored highest for grazing areas. Although both easements and profits scored highest, they both scored lowest on legal acceptability regarding their implementation in the pastoral areas. The low scores on legal acceptability are perhaps because the migration corridors and grazing areas fall under two distinct tenure systems: formal and customary. Exclusive rights on formal land (group ranches) annulled customary land rights. Introducing a new right of easement or profit may be challenging as it entails interference with the private rights on the formal land. On customary land, the pastoralists' rights on migration corridors and grazing areas remain unhindered. Moreover, legal requirements for introducing easements and profit on customary land are not addressed in the laws.

Meanwhile, open access scored as the second best tenure option for both migration corridors and grazing areas, although their total scores are just a few points higher than the baseline. This position is due in large part to the relatively good score for the equity criterion, reflecting the fact that the option is nondiscriminatory because it does not exclude others from access to the land. The baseline (practices as they presently exist) was viewed as the third best option for securing access to migration corridors and grazing areas. This suggests that if this option was chosen there would be no change in the land tenures, land use, and management practices.

Context 2: Dry Season Grazing in Non-Pastoralist Areas

Before scoring the tenure options suitable for pastoralists' land rights in non-pastoral areas, the land professionals were asked for their views on whether pastoralists should be allowed to migrate with their livestock and access dry season resources in the non-pastoralist areas. Most respondents consider that pastoralist migrations ought not to be allowed in farming areas but could be allowed on government lands: wildlife parks and forests. All 13 land professionals responded to this question. The majority of the respondents (12 out of 13) indicated that pastoralists should not be allowed access to these areas at any time of the year, and only during required dry seasons.

Moving beyond the question of whether migration should be allowed in non-pastoralist areas, a varying proportion of the 13 respondents (between 7 and 10) answered further sections. Those not responding argued that pastoralists' land rights should not seasonally overlap with private tenures. They argued that it is cumbersome to coordinate overlapping land rights especially where different land uses are exercised and that wildlife needs to be protected and conserved and is a very important source of revenue for the government; some emphasized the need to protect and conserve forests because opening them up by allowing migrations and grazing would provide an avenue for their exploitation and destruction. Those responding believed that the seasonal migration and grazing in non-pastoralists areas could be coordinated. Tenure options preferred by the land professionals responding are presented below.

With regard to the specific case of farming areas, nonownership forms of tenure are the preferred tenure options for supporting pastoral tenures. Reserved land, in this case land that the government would provide specifically for pastoral purposes (migrations and grazing), scores best overall for supporting pastoralists' tenures in the farming areas. Second preferred option is negotiation, a process aimed at leading to an agreement between pastoralists and farmers. Nonownership forms of tenure, that is, easements to support migration corridors and grazing areas score fairly positive by taking the third position. The fact that is common between these nonownership forms of tenure is that they score negatively in administrative feasibility, technical feasibility, and legal acceptability if they were to be implemented in farming areas. This means interfering with private rights of the "legal right holders" by deducting their rights and giving them to pastoralists in specified dry seasons. A critical evaluation of the ease of introducing easement and profit (administrative criteria), of judgment of the resources and skills needed to implement them (technical feasibility), and of the legal acceptability of implementing these tenure options in the farming areas may be expected to be challenging, considering that private tenures (individual or government) are the dominant tenure option in these areas.

Regarding the specific cases of wildlife parks and forests, nonownership forms of tenure are preferred for supporting pastoral tenures, just as in farming areas. Reserved land also consistently received high scores—for supporting migration corridors and for supporting dry season grazing. These results suggest that special areas within wildlife parks and forests could be designated for use by pastoralists. The nonownership rights of easement and profit are the second best options, whereas the negotiation option received the second best position.

Ownership forms of tenure, that is, freeholds and leaseholds, are dismissed as tenure options for supporting pastoral tenures in non-pastoral areas as they score negatively in all criteria.

Discussion

The purpose of this study was to find out what tenure options that have statutory underpinnings have the potential to secure pastoralists' seasonal land rights in a dynamic cadastre. The context of the study is northern Kenya, where the interests of pastoralists' land rights seasonally coincide with private land tenures—individual and government tenures. Land professionals in Kenya subjectively evaluated and scored the tenure options according to their potential to support the seasonal land rights on migration corridors and dry season grazing areas. This section discusses the implications and relevance of the proposed tenure options on pastoralists' dynamic land rights.

Context 1: Tenure Options for Securing Pastoralists' Land Rights within Pastoral Areas

Within the pastoralists' areas, easement, a nonownership form of tenure, scored highest as having potential to secure pastoralists' right-of-way on the migration corridors. According to the laws regulating land, the land needs to be owned by a different party for it to be legally possible to create an easement. Gray and Gray (2005) explain the four requirements that need to be fulfilled to create rights of easements: (1) every easement is linked to two parcels of land, its benefit being attached to a "dominant tenement" and its burden being asserted against a "servient tenement"; (2) land parcels should be close to one another; (3) the dominant and the servient parcels should be owned by different persons; and (4) easements must be created through express grant, implication or prescription. As such, group ranches, which are statutorily regulated, fulfill the requirements for creation of easements. Although easements are often used between neighboring parcels, pastoralists' migration corridors run across group ranches, community land, and beyond. In such a case, as soon as easements on statutory tenures (group ranches) have been navigated across, the pastoralists join the migration corridors in the community land, which are managed under customary law.

We are unaware of the real-life application of easements across group ranches to secure migration corridors in the formal land administration system. Public roads to support the right-of-way during pastoralists' seasonal migrations are supported by examples from a number of countries such as Niger, Australia, Burkina Faso, Norway, Mali, and other countries (South Australia 1989; Northern Territory of Australia 2011; Burkina Faso 2002; Mali 2001; Niger 2008; Norway 2007). The users' public roads for livestock migrations have only use rights in the said countries. Use rights during livestock migrations are aligned to their migration calendars. The Niger pastoral code, for example, even takes into consideration the local contexts when determining the requirements for minimum widths of the migration corridors and also provides for enforceability through a structure of penalties for obstructing the corridors during the period that pastoral rights apply (Niger 2008).

On securing pastoralists' land rights to dry season grazing areas, profit, a nonownership form of tenure, scored the highest. Similar to the creation of easements, land must be held by another party for a right of profit to be established. This means that, like easements, group ranches fulfill the requirements for the rights of profit to be created. Theoretically, the right of profit on group ranches may be possible. Practical applicability of the right of profit on group ranches might however be challenged by the nature in which the group ranches are held, that is, joint ownership, and how they are used, that is, communal use.

Context 2: Tenure Options for Securing Pastoralists' Land Rights in Non-Pastoral Areas

Reserved land as the preferred tenure option for securing pastoralists' right-of-way on migration corridors in farming areas, wildlife parks, and forests is evident. This is similar for securing pastoralists' rights to graze in those areas. Subject to any other written law, the Kenya Government Lands Act (Kenya 2010b) (§35) allows the registration of limited rights on "un-alienated Government land" to be reserved for special purposes. Un-alienated government land means, "Government land which is not for the time being leased to any other person, or in respect of which the Commissioner has not issued any letter of allotment" (Kenya 2010b). Examples of government land include areas reserved for a township, farmlands, and the like. Reserved land, in this case, is land that the government would provide specifically for pastoral purposes—right-of-way on migration corridors and right to grazing.

The use of reserved land for pastoral purposes is supported by examples from a number of countries. In Niger, reserved lands are allocated for the purpose of grazing. They are classified as the public or private domain of the state, or of a territorial reserve for grazing or pastoral development (Niger 2008). In Mongolia, reserved areas for grazing during spring, summer, autumn, and winter are categorized and, in addition, some reserved lands are secured in the event of natural disasters, such as droughts (Mongolia 2002). The government of Mongolia determines the extent of the reserved lands, including their boundaries, and the limitations on their use. In Norway, reindeer herders have designated areas reserved for grazing and grazing rights are aligned to grazing seasons, that is, spring, summer, autumn, and winter (Norway 2007). In Burkina Faso (2002), local communities and local authorities are involved in establishing local regulations on access to land reserved for grazing for the purpose of sustainable management of these resources. From these examples, access to reserved lands is aligned to grazing seasons/calendars and across the geographical landscapes to which pastoralists' land rights apply. These examples show not only that reserved lands may serve as an option for securing and preserving dry season grazing areas but also that local communities and the government can be involved in laying the basis for ecologically and economically sustainable utilization of grazing resources. In these countries, the users of reserved land do not have the rights to control the land, but they have limited access and use rights crucial for pastoralists.

Negotiations were considered to be the second best option after reserved. This result is in line with practices in countries such as Burkina Faso and Guinea. In these countries, pastoralists' migration corridors to farming areas are recognized and secured in the pastoral codes and the main means by which pastoralists gain access to private land are negotiations with farmers (Burkina Faso 2002; Guinea 1995). Although experience from West Africa show that negotiations with holders of private rights are slow to produce agreements, and sometimes even slower to produce results on the ground (Nianogo and Thomas 2004), negotiations may be perceived as being efficient when parties discuss access to private land, therefore keeping conflict to a minimum. However, access to the private property cannot be guaranteed because it is dependent on the landowner's permission.

The results of this study give insights into what tenure options that have statutory underpinnings have the potential to secure pastoralists' seasonal land rights in the formal land administration system. As much as nonownership forms of tenure are important to introduce the dynamism required to manage pastoralists' seasonal land rights, we take into account that the methodology used in this study contains a number of limitations. First, the scoring of tenure options and use of closed-ended questions led to obtaining purely subjective opinions from the land professionals. Any qualitative information about why they made those tenure choices is not captured. Second, the criteria used in the methodology for general policy analysis issues—Bardach (2009) and Patton and Sawicki (1986)—do not include things such as pasture availability or things that directly influence pastoralism. The methodology, which was used in the context of evaluating tenure options, produced results that mirror the practices happening in other countries to secure pastoralists' tenures. An example can be derived from reserved land, which was the most preferred and recurrent tenure option for supporting pastoralists' dynamic land rights. Although the successes and weaknesses of these approaches in achieving the desired policy goals may need to be quantified, the pastoral legislation in the West African subregion has brought about some positive developments, such as the recognition of the economic importance of livestock rearing, reinstatement of pastoralism as a productive land use, preservation of pastoral mobility, opportunities for herders to gain access to required resources, and reinstatement of indigenous methods of conflict resolution (Touré 2004).

Conclusions

This study started with the well-known notion that parcel-based tenures in the formal land administration system introduce territorial fixity and are unable to support pastoralists' spatial–temporal land rights. This study sought the views of land professionals regarding what tenure options with statutory underpinnings have the potential to secure pastoralists' seasonal land rights in a dynamic cadastre.

The results showed that nonownership forms of tenure such as easements, profit, or use rights on reserved land have the potential to support pastoralists' dynamic land rights. Specifically, easements are applicable on group ranches to secure migration corridors. Profit also applies on the group ranches to secure dry season grazing areas. Group ranches, being statutory tenures, fulfill the requirements for the creation of these nonownership rights. Rights-of-way on migration corridors and access to dry season grazing can continue unchanged where customary tenures prevail. Also, results showed that reserved lands for pastoral purposes (both migration corridors and grazing lands) with use rights aligned with pastoralists' seasonal land rights were considered appropriate. Arguably, these are important findings in the context of tenure privatization and policies that have been promoted by governments to support sedentary land uses.

This study proposes timely identification and documentation of migration corridors and dry season grazing areas, as availability of such information during adjudication and demarcation could support informed decision making regarding relevant tenure options prior to the allocation of exclusive rights. In the context of this study, existence of pastoralists' migration corridors and calendars while evaluating the tenure options might have led the land professionals to conclude that nonownership tenures are appropriate for pastoralists' seasonal land rights within the statutory tenure system.

This study has the potential to lead to a better understanding of tenure options that could support pastoralists' spatial–temporal land rights in a dynamic cadastre. Areas of further research are suggested by the study: investigation on the feasibility of nonownership forms of tenure for pastoralists' land rights—as the corridors and grazing areas could change as influenced by various biotic and abiotic factors. In addition, the work could focus more on qualitative approaches to explore links between pastoralists' seasonal land rights in non-pastoralist areas, to reveal the informal rules, rights, and restrictions created by herders and private land use actors. This may provide clues to the nature of temporary arrangements regarding reciprocal use of land, which enables the informal coexistence of pastoral land rights and private land rights within the formal land administration system. Further, quantitative studies might provide clues on the effects of livestock densities/carrying capacity and intensity of use on the formal migration corridors.

References

Awogbade, M. O. 1987. Grazing reserves in Nigeria. *Nomadic Peoples* 23:19–30.

Bardach, E. 2009. A Practical Guide for *Policy Analysis*: The Eightforld Path to More Effective Problem Solving. Washington, DC: CQ Press.

Blench, R. M. 2001. *Pastoralism in the New Millennium (Animal Health and Production Series, 150)*. Rome, Italy: FAO.

Burkina Faso. 2002. Loi n°034-2002/an portant loi d'orientation relative au pastoralisme au Burkina Faso, accessed April, 2015, http://faolex.fao.org/cgi-bin/faolex.exe?rec_id=032528&database=faolex&search_type=link&table=result&lang=eng&format_name=@ERALL.

Coldham, F., and R. Simon. 1979. Land-tenure reform in Kenya: The limits of law. *Journal of Modern African Studies* 17 (4):615–627. doi: 10.2307/160742.

Dale, P. F., and J. McLaughlin. 2000. *Land Administration*. Vol. 9, *Spatial Information Systems and Geostatistics Series*. Oxford, United Kingdom: Oxford University Press (OUP).

FIG 1995. Statement on the Cadastre. International Federation of Surveyors, accessed May 4, 2015, https://www.fig.net/commission7/reports/cadastre/statement_on_cadastre.html.

Galaty, G. D. 1988. Scale, Politics and co-operation in organization for East African Development. In: Attwood D. W. and Baviskar B. S. (Eds), *Who shares cooperatives and rural development?* Oxford University Press.

Gray, K. J., and S. F. Gray. 2005. *Elements of Land Law.* Oxford, United Kingdom: Oxford University Press.

Guinea. 1995. *Loi L/95/51/CTRN Portant Code Pastoral 1995—Guinea*. Accessed June. Retrieved fom http://faolex.fao.org/docs/pdf/gui4469.pdf.

Kenya. 1963. *Registered Land Act Cap. 300 of the Laws of Kenya*, accessed April, 2015, http://kenyalaw.org/kl/fileadmin/pdfdownloads/Acts/RegisteredLandActCap300.pdf.

Kenya. 2009. *National Land Policy*. Ministry of Lands, Republic of Kenya.

Kenya. 2010a. *The Constitution of Kenya.*, accessed April, 2015, http://www.kenyalaw.org:8181/exist/kenyalex/actview.xql?actid=Const2010.

Kenya. 2010b. *The Government Lands Act*, Chapter 280, accessed April, 2015, http://kenyalaw.org/kl/fileadmin/pdfdownloads/Acts/GovernmentLandsActCap280.pdf.

Kenya. 2013. *Community Land Bill.*, accessed April, 2015, http://kenyalaw.org/kl/fileadmin/pdfdownloads/bills/2014/KenyaGazetteSupplementNo147.pdf.

Kimani, K., and J. Pickard. (1998). Recent trends and implications of group ranch subdivision and fragmentation in Kajiado District, Kenya. *Geographical Journal* 164:202–213.

Lemmen, C. H. J. 2012. *A Domain Model for Land Administration*. PhD thesis, Delft, The Netherlands: NCG.

Lengoiboni, M., A.K. Bregt, and P. van der Molen. 2010. Pastoralism within land administration in Kenya: The missing link. *Land Use Policy* 27 (2):579–588.

Lengoiboni, M., P. van der Molen, and A. K. Bregt. 2011. Pastoralism within the cadastral system: Seasonal interactions and access agreements between pastoralists and non-pastoralists in Northern Kenya. *Journal of Arid Environments* 75 (5):477–486.

Liao, C., S. J. Morreale, K.-A. S. Kassam, P. J. Sullivan, and D. Fei. 2014. Following the Green: Coupled pastoral migration and vegetation dynamics in the Altay and Tianshan Mountains of Xinjiang, China. *Applied Geography* 46 (0):61–70. doi: http://dx.doi.org/10.1016/j.apgeog.2013.10.010.

Mali. 2001. *Loi N 01-004 DU 2 7 FEV. 2001—Portant Charte Pastorale Du Mali*, accessed April, 2015, http://faolex.fao.org/docs/texts/mli25376.doc.

McCarthy, N., B. Swallow, M. Kirk, and P. Hazell. 1999. *Property Rights, Risk, and Livestock Development in Africa*. Washington, DC: International Food Policy Research Institute.

Mongolia. 2002. *Law of Mongolia on Land*, accessed April, 2015, http://faolex.fao.org/docs/texts/mon62064.doc.

Mwangi, E. 2007. Subdividing the commons: Distributional conflict in the transition from collective to individual property rights in Kenya's Maasailand. *World Development* 35 (5):815–834. doi: http://dx.doi.org/10.1016/j.worlddev.2006.09.012.

Niamir-Fuller, M. 2005. Managing mobility in African rangelands. In Mwangi E. (Ed), *Collective Action and Property Rights for Sustainable Rangeland Management*. CAPRi

Nianogo, A., I. Thomas. 2004. *Forest-livestock interactions in West Africa*. Lessons learnt on sustainable forest management in Africa. KSLA/AFORNET/AAS/FAO report.

Niger. 2008. *Republic of Niger.* Pastoral draft law version March 2008.

Nori, M. 2007. *Mobile Livelihoods, Patchy Resources and Shifting Rights: Approaching Pastoral Territories*. Rome, Italy: International Land Coalition.

Northern Territory of Australia. 2011. Pastoral Land Act, accessed April, 2015, http://faolex.fao.org/docs/pdf/nt18581.pdf.

Norway. 2007. *Sami Reindeer Act LAW 2007-06-15 nr 40*, accessed April, 2015, http://faolex.fao.org/docs/texts/nor77640.doc.

Okoth-Ogendo, H. W. O., and W. Oluoch-Kosura. 1995. *Final Report on Land Tenure and Agricultural Development in Kenya*. Nairobi, Kenya: Ministry of Agriculture, Livestock Development and Marketing.

Patton, C. V., and D. S. Sawicki. 1986. *Basic Methods of Policy Analysis and Planning*: Englewood Cliffs, NJ: Prentice-Hall.

Pugh, S. 1996. *Creating Innovative Products Using Total Design: The Living Legacy of Stuart Pugh,* Don Clausing and Ron Andrade (Eds). Boston, USA: Addison Wesley Longman.

Rietbergen-McCracken, J., and H. Abaza. 2014. *Economic Instruments for Environmental Management: A Worldwide Compendium of Case Studies*: London, United Kingdom: Routledge.

Rutten, M. M. E. M. 1992. *Selling Wealth to Buy Poverty: The Process of the Individualization of Landownership Among the Maasai Pastoralists of Kajiado District, Kenya, 1890–1990*: Breitenbach, Saarbrücken, Germany: Verlag breitenbach Publishers.

South Australia. 1989. *Pastoral Land Management and Conservation Act*, accessed April 2015, http://www.legislation.sa.gov.au/LZ/C/A/PASTORAL%20LAND%20MANAGEMENT%20AND%20CONSERVATION%20ACT%201989.aspx

van der Molen, P. 2003. *The Future Cadastres—Cadastre after 2014*. FIG Working Week 2003. Paris, France, April 13–17.

Western, D., R. Groom, and J. Worden. 2009. The impact of subdivision and sedentarization of pastoral lands on wildlife in an African savanna ecosystem. *Biological Conservation* 142 (11):2538–2546. doi: http://dx.doi.org/10.1016/j.biocon.2009.05.025.

Section IV

Measuring the Impacts

12

Land Administration Impacts on Land Use Change

Peter Fosudo, Rohan M. Bennett, and Jaap Zevenbergen

CONTENTS

Introduction

Conventional land administration attempts to formalize the interaction between people, the state, and the landscape (Farley et al. 2012; Gerlak 2014; Wannasai and Shrestha 2008). Land tenure formalization, also known as land tenure regularization (LTR), is the process that seeks to create state recognition of land rights, and subsequently land tenure security, in places where it previously did not exist (Williamson et al. 2010). LTR articulates the state-sanctioned bundle of rights, restrictions, and responsibilities that relates to the land (Durand-Lasserve and Selod 2009; Zevenbergen 2002). The desired outcome of LTR is that land users can more assuredly make decisions about land: the landscape can be reshaped by new land users and uses stemming from unambiguous land tenure security (Farley et al. 2012; Wannasai and Shrestha 2008).

That LTR will impact a landscape is generally agreed upon; however, the extent of the impact is less clear and highly dependent on the context. This is where the study area of land administration overlaps with the related area of land use change (LUC). Studies on the changing relationship between people and the environment are essential (Veldkamp and Lambinb 2001). It is also important to understand the impact of interventions in terms of landscape outcomes, changes to the built environment, as well as the multivariate natures of the change (Hersperger and Burgi 2010; Orenstein et al. 2011).

In this regard, works focus on measuring the impact of agrarian reforms (Farley et al. 2012), understanding the causes of LUC using statistical analysis (Serneels and Lambin 2001), and determining the demographic and socio-economic drivers of LUC (Mena et al. 2006; Mottet et al. 2006). The techniques embedded in geographic information system (GIS) and remote sensing tools underpin much of the work.

Despite much research on LUC, limited works appear to focus on how tenure formalization programs impact LUC. Of the studies that do exist, many tend to use only conventional remote sensing techniques and algorithms to assess changes in a Spatiotemporal manner: the social aspect is either not considered or is studied separately, meaning a more holistic viewpoint is lacking. This is particularly important in peri-urban areas, the transitional zones between urban and rural areas, where LUC can be rapid and a multitude of factors drive the process (Arko-Adjei 2011). Existing land users, often agriculturalists, seldom possess secure land tenure, and administrative voids may result in limited controls on LUCs. The impacts of LTR are not always holistically assessed in these contexts.

This gap in knowledge relating to the impact of LTR on agricultural LUC in peri-urban areas provides the focus for this chapter. Specifically, the effect of the Rwandan LTR program on agricultural LUC is examined over a specific epoch. An assessment of spatial, temporal, and social changes between 2008 and 2013 in an area of Kigali, Rwanda, is made. The types of rights held before the LTR program, nature of the rights held following LTR, and impact that the rights had on the decision making of landowners are studied. The actor-change (A-C) theory of Hersperger et al. (2010) acts as the basis for understanding LUC, and the spatiotemporal analysis makes use of GIS.

The chapter seeks to reveal a new understanding of the relationship between LTR programs and LUC through the application of a mixed socio-spatial method. The work will help to enhance understanding of the way in which people react to LTR policies in peri-urban areas. It will demonstrate to policy makers the potential intermediate and long-term consequences of such programs. The chapter first provides a theoretical background on LUC before detailing the research methodology, results, and key discussion points. The chapter concludes by hypothesizing future research directions.

Theoretical Perspective

LUC is an extension of the term "land use," a term with generalized understandings but multifaceted meanings. Many technical definitions might not consider the role of people. For Turner II et al. (1994), land use is as follows:

> the biophysical state of the earth's surface and immediate subsurface.

However, others make explicit mention of human involvement. Briassoulis (2000), in line with the Food and Agriculture Organization (FAO) (1995) suggest as follows:

> ... land use concerns the function or purpose for which the land is used by the local human population and can be defined as the human activities which are directly related to land, making use of its resources or having an impact on them.

In a similar vein, Sultana and Powell (2010) see land use as "the way the surface of the earth has been used through human activities like transport, farming, and industry." It can be viewed as "how" and "why" the land and its resources are being influenced (Briassoulis 2000; Meyer and Turner II 1994). In this regard, different people-made classifications emerge to understand the how and why: residential (for shelter/housing), commercial (for trade and commerce), agricultural (for farming), institutional, and recreational.

Following on, LUC can be described as the people-driven actions that modify how the Earth's surface is utilized. Meyer and Turner II (1994) view LUC as change of land use from one type to another, or intensification of the existing land use. Put more simply, LUC may refer to the conversion or modification of land (Briassoulis 2000). The change can be quantified in terms of coverage (Briassoulis 2000).

LUC is often viewed as a local environmental activity (Abiodun et al. 2011; Foley et al. 2005, 570). However, despite differences in land use practices globally, the consequences of LUC remain similar, if not connected. When the results of LUC are aggregated globally, they contribute to the changing carbon cycle, the loss of total agricultural land, the loss of biodiversity, habitat fragmentation, and ecological overexploitation (Foley et al. 2005; Houghton et al. 2001; Pielke et al. 2002). On the specific case of peri-urban LUCs, high competition for land, migration from rural to urban areas, natural population growth, and socioeconomic systems operating at various scales influence the changes (Dubovyk et al. 2011). Doygun (2009) observes that the agricultural land use in close proximity to urban areas is more affected by such changes (either conversion or modification).

Meanwhile, it is also necessary to define the relationship between land use and "land cover." At times they are used interchangeably, but at other times they are distinguished (Briassoulis 2000; Li et al. 2005; Muttitanon and Tripathi 2005; Orenstein et al. 2011). Briassoulis (2000) describes land cover as

> the physical state of the land surface: as in cropland, mountains, or forests.

Moser (1996) and Meyer and Turner II (1994) add that

> it embraces ... the quantity and type of surface vegetation, water, and earth materials.

Invariably,

> The term originally referred to the type of vegetation that covered the land surface, but has broadened subsequently to include human structures, such as buildings or pavement, and other aspects of the physical environment, such as soils, biodiversity, and surfaces and groundwater (Briassoulis 2000; Meyer and Turner II 1994; Moser 1996).

The close relationship in the way that the terms are used brings about their interchangeable connections, which makes Meyer and Turner II (1994) posit that

> A single land use may equally correspond fairly well to a single land cover.

However, a particular land cover may be attributed to different uses, for instance, agricultural land with varieties of crops or forestland with several purposes. In terms of administrative or spatial delineation, a single land use can accommodate several smaller land cover units, for example, woodlands, settlements, and pastures, among others (Briassoulis 2000). At any rate, the often synonymous understanding of the terms leads to the use of the term "land use/land cover." In the work by Lu et al. (2004), land use/land cover changes are categorized as follows: long-term natural changes in climate conditions, geomorphologic and ecological processes such as soil erosion and vegetation succession, human-induced alterations of vegetation cover and landscapes such as deforestation and land degradation, interannual climate variability, and the greenhouse effect caused by human activities.

Regarding scientific research on LUC, numerous perspectives are evident. Farley et al. (2012) address the issue from the perspective of impact on agrarian reform. Serneels and Lambin (2001) focus on developing statistical analyses to create a general understanding of LUC. Meanwhile, Mena et al. (2006), Mottet et al. (2006), and Farley et al. (2012) determine the impact of demographic and socioeconomic drivers, and geo-biophysical factors, to enhance the understanding of LUC causes and effects. LUC models are another area of scientific enquiry (e.g., Verburg et al. 2004). These works provide the theoretical underpinnings for this chapter.

Existing LUC models are generally based on (1) driving forces, (2) cross-scale dynamics, (3) level of analysis, (4) spatial interaction and neighborhood effects, (5) temporal dynamics, and (6) level of integration (Verburg et al. 2004). Another model is the one proposed by Hersperger et al. (2010). Components of the LUC model are identified as driving forces, actors, and change. Driving forces, along with actors, shape the changes in land use. They are a complex system of interdependent interactions occurring across space and time. Actors are decision makers and might be individuals, agencies, and institutions. They can affect the driving force. From this, four change types are derived: (1) driving force-change (DF-C), where the driving force causes the change; (2) driving force-actor-change (DF-A-C), where multiple driving forces influence an actor to make an LUC; (3) driving force actor-change (DFA-C), where there is an interplay between the driving force and the actor; and (4) A-C, where an actor's reasoning and values influence the land use causing change.

In this chapter, the A-C model is adopted; as per suggestions from Briassoulis (2000), the model's characteristics were found to fit the intended case study context, including geographical extent, volume of land use, and data availability. This can be used to model the type and extent of LUC arising from actor decisions following LTR. The selection is further justified in the section "Methodology."

Methodology

To determine the relationship between LTR programs and LUC, an analytical method considering social, spatial, and temporal aspects was applied to a case study area in the Rwandan capital of Kigali. The study areas, Kinyaga and Masoro cells, were two different cells within Bumbogo and Ndera of Gasabo districts of Kigali. Like much of Rwanda, the terrain is hilly and the general land use in the area was previously agricultural. However, urbanization processes are evident: residential, commercial, and industrial uses appear on the landscape. LTR was conducted in the area after 2008.

At its root, the analytical method was based on Hersperger et al. (2010). The A-C model was adopted. This model was considered appropriate because: (1) it was considered that reasoning and values of the actor constituted the major influencing factor on LUC following LTR; (2) the study area's geographical extent was considerable—two cells in two different sectors were included; (3) the volume of land uses and owners, parcels, land size, and owners could be given proper consideration during data collection process; and (4) consideration could also be given to actor behavior and land change data. Regarding the A-C model, the following were identified for the case study: (1) the driving force was considered to be the secure tenure resulting from LTR; (2) actor was considered to be the landowner; and (3) change relates to agricultural land use transitioning to other uses including residential, industrial, and educational uses.

Regarding data collection, the study adopted a dual sampling technique. This included stratified and random sampling of the study area. Two strata from Gasabo districts, namely, Bumbogo and Ndera sectors, were used. Meanwhile, two cells Kinyaga and Masoro, one cell from each of the two sectors of Bumbogo and Ndera, were also used to ensure representation. Data collection was both primary and secondary in nature.

Primary data were gathered through direct interviews of 25 randomly selected landowners in the different cells in September/October 2013. This data were used to determine (1) the land tenure system in the area, (2) whether landowners or occupants possessed land certificates, and (3) whether the possession of land certificates impacted land use decision making. In addition, government officials from the Rwandan Natural Resources Authority (RNRA) Land and Mapping Department were interviewed. This assisted in determining if the government played a role in LUC in the area. Parallel to the interviews, an inventory of land uses in the study area was undertaken. Classifications for each parcel used included the following: built, partly built, and vacant.

Secondary data were collected from various sources. A raster image of the Gasabo district, on July 7, 2013 (i.e., post LTR), was collected from the Remote Sensing Laboratory of the Faculty of Geo-Information Science, ITC, University of Twente, Enschede, the Netherlands. In addition, a raster data set (i.e., a pre-LTR orthophoto map of 2008) of the study cells was obtained from the RNRA Land and Mapping Department. Vector data of the study cells (shape files) were made available by the same agency. Other secondary data included land documents, literature, journal articles, the Rwandan land policy, and related laws.

To evaluate the change relationship between LTR and LUC, two approaches were utilized. First, using the interview data the relationship between LTR outputs (land titles) and LUC was determined. For each interview, an influence rank of 1–3 was allocated Abushnaf (2013): the significance of ownership certificates influencing land use decisions could be derived for the study area. Second, from the imagery and spatial data change detection was performed between 2008 and 2013. The aim of this change detection was to determine the extent of LUC from 2008 to 2013. Specifically, agricultural land use converting to other forms was the focus. This involved change detection, which meant applying multi-temporal data sets to analyze temporal effects of the phenomenon (Singh 1989). Using ArcGIS, the parcel shape files resulting from LTR were overlaid atop the two raster images from 2008 and 2013. For each parcel, a different land use category (built, partly built, and vacant agricultural lands) was assigned. From the analysis, the size and percentage of each land use could be calculated for each epoch. By synthesizing the results of the qualitative social data (interviews) and Spatiotemporal approach (GIS data), an overarching understanding of the impact of LTR on LUC for the peri-urban area could be ascertained.

Results

Results from the enquiry are presented in the following way. First, the status of changing land tenures and its impact on land use decision making, as revealed by interviewees in the case study area, are revealed. Second, the changes in land use as ascertained by the spatial analysis are presented.

From the interview data analysis, the changing status of land tenure and its impact on land use decision making could be investigated. Prior to LTR, land tenures in the study area were found to be customary in nature and fell into one of four categories: *Ubukonde, Igikingi, Inkungu,* and *Gukeba* (Rurangwa 2013). Following LTR, of the 25 respondents 20 indicated that they now held an LTR land certificate and 2 suggested that they possessed a customary tenure, whereas 3 did not know what kind of tenure they held. Although LTR appears to have been pervasive, it can be said that the remnants of former systems and understandings persist. Meanwhile, with respect to how the bestowed LTR land titles impacted land use decisions, 18 of the 25 respondents were motivated to invest in, or further develop, their lands; 4 suggested that the titles did not influence decision making; and 3 respondents were indifferent to the question.

From the spatial data analysis (Figure 12.1), the amount of agricultural LUC over the two epochs could be determined. In 2008, the two cells Kinyaga and Masoro were revealed to contain 1597 and 1965 each, and constituted 515 ha and 630 ha, respectively. Therefore, in total the study area consisted of 3562 parcels and 1144 ha. Prior to LTR, 1123 parcels (67.1 ha) were considered built, 365 parcels (619 ha) were partially built, and 2074 parcels (458 ha) were agricultural. This meant that in 2008, 58% of land parcels in the area, or 40% of the total study area, were considered agricultural. Subsequent to LTR in 2013, 2001 parcels (332 ha) were considered built, 481 parcels (457 ha) were considered partially built, and 1080 parcels (354.5 ha) remained as vacant agricultural parcels. The number of agricultural parcels had dropped from 58% to 30% (28% change) of the total number of parcels, and the total area of agricultural land use had dropped from 40% to 31% (9% change) within the 5 years (Figure 12.2).

Discussion

This chapter seeks to reveal the relationship between LTR programs and LUC. The subsequent discussions explore the extent of agricultural LUCs following LTR, whether those LUCs were driven by landowners motivated by LTR, and the success of the method applied.

LUC is certainly observed in the study area between 2008 and 2013. The number of built parcels rose by 25% during the period. This constitutes 23% of the total land area. In addition, whereas 365 parcels (619 ha) were already in transition in 2008 (i.e., partially built), an even greater number (481 parcels or 458 ha) was in transition in 2013. Of course, some may have been in transition for the entire study period; however, a certain level of LUC cannot be denied. In the study area, it can be concluded that agricultural land uses were shifting to industrial, commercial, or

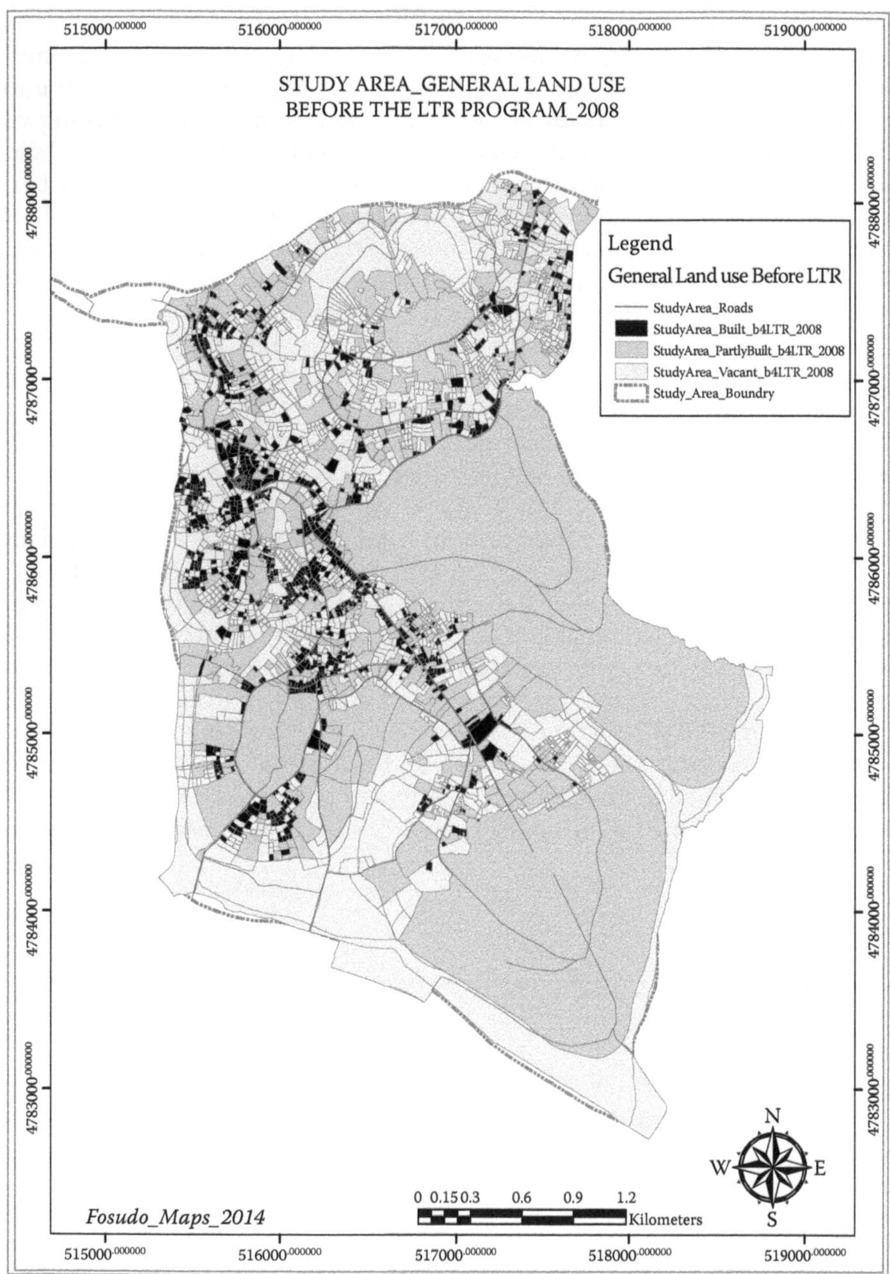

FIGURE 12.1
Land uses in the study area in 2008.

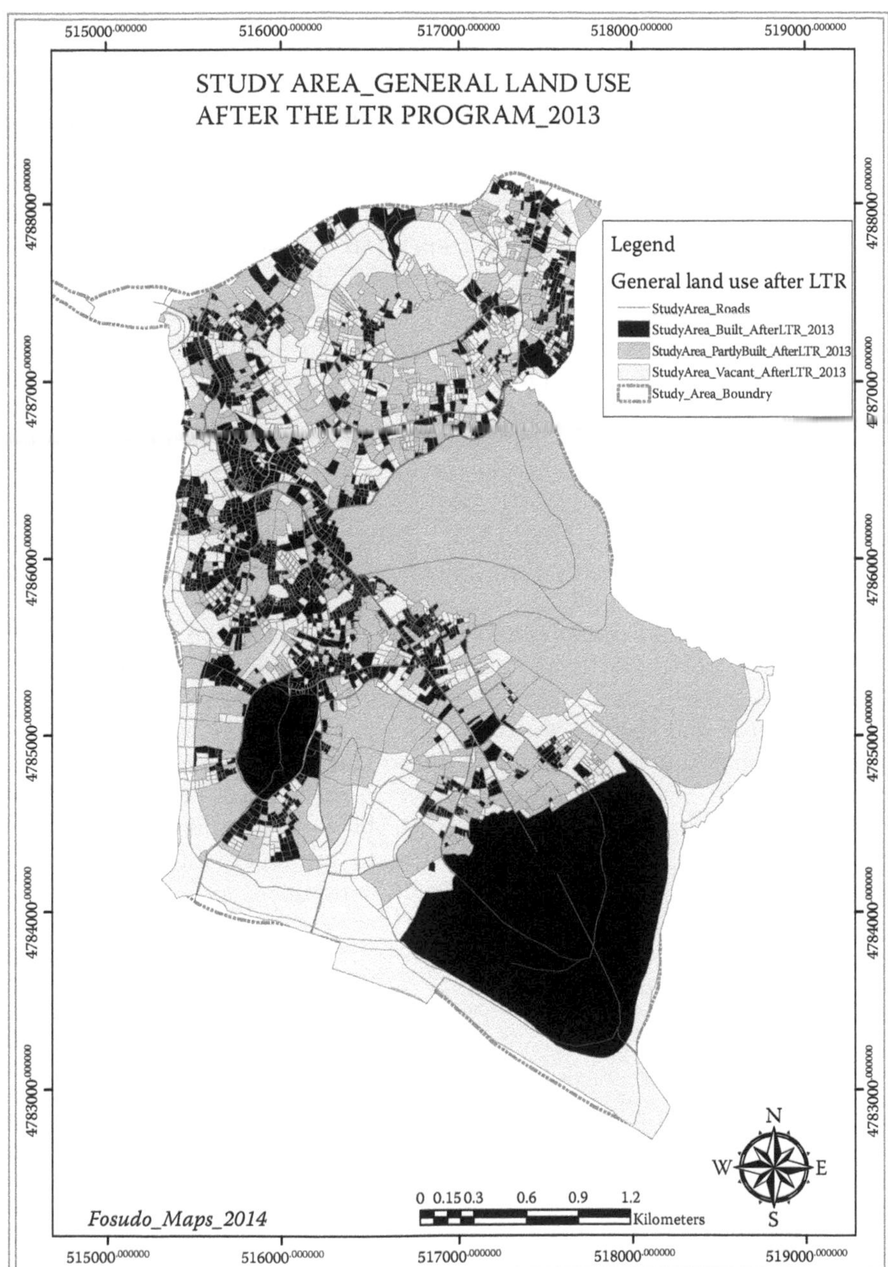

FIGURE 12.2
Land uses in the study area in 2013.

residential uses. The significance of LUC is difficult to judge as no control group existed.

Determining how much of the LUC can be attributed to LTR is a difficult task. Although LTR might have had a significant influence, there could have been other contributing factors. The qualitative interview data provide some insight in this regard: 18 of the 25 respondents suggested that the outputs of LTR, that is, the land titles, motivated their actions in the period between 2008 and 2013, particularly with respect to investment. Unfortunately, a control group was not available and, in addition, those who may have been owners in 2008, and had subsequently sold lands on account of LTR, were not directly taken into account. At any rate, it appears national land policies, laws, and the subsequent LTR program had some level of impact on agricultural LUCs in the study area. A deeper study would be needed to distinguish and disaggregate other possible influences, for example, the creation and implementation of city master plans, rural to urban migration, and economic growth more generally. Overall, based on the compiled results, it does appear that the agricultural LUC can be traced to LTR: LTR, via secure tenure, motivated landowners to act, and land uses subsequently changed based on these decisions.

Regarding the applied methodology, the change detection method, based on multi-temporal spatial data sets and a simple classification system, helped to derive a discrete understanding of LUC in terms of land parcels and areas. A limitation is that only vector data sets from 2008 were available: parcel mutations including subdivision and consolidation, if they occurred, could not be observed. The parcel mutations would reveal more about the nature of LUCs that occurred. The supporting interview data could be used to discern reasons for the changes; however, a greater number of interviews would have made the accompanying reasoning more robust.

Conclusion

The chapter sought to reveal a new understanding of the relationship between LTR programs and agricultural LUC in peri-urban areas through the application of a mixed socio-spatial method. The approach derived inspiration from the A-C model of Hersperger et al. (2010): it was perceived that the LUC occurring was driven by actors. The results revealed that LUC occurred in the study area over the study period: aerial imagery and high-resolution satellite imagery can indeed be used to detect LUC in peri-urban areas. However, the extent to which the LUC can be attributed to LTR is still open to debate. Certainly, the qualitative interview data revealed that LTR influenced the majority of the sample population; but other factors may have also been attributable, not the least being city master plans, rural and urban migration, and sustained economic growth more generally. Further research could attempt to distinguish the influence of LTR from other interventions and driving forces. Overall, it is felt that the work helps to enhance understanding of the way in which people react to LTR programs in peri-urban areas. It demonstrates to policy makers the potential intermediate and long-term consequences of such programs.

References

Abiodun, O. E., Olaleye, J. B., Dokai, A. N., and Odunaiya, A. K. 2011. *Land Use Change Analyses in Lagos State From 1984 to 2005*. FIG Working Week, May 18–12, Marrakech, Morocco.

Abushnaf, F. F., Spence, K. J., and Rotherham, I. D. 2013. Developing a land evaluation model for the Benghazi region in Northeast Libya using a geographic information system and multi-criteria analysis. *APCBEE Procedia, 5* (0), 69–75. doi: http://dx.doi.org/10.1016/j.apcbee.2013.05.013.

Arko-Adjei, A. 2011. *Adapting Land Administration to the Institutional Framework of Customary Tenure: The Case of Peri-Urban Ghana*. Amsterdam, The Netherlands: IOS Press BV.

Briassoulis, H. 2000. *Analysis of Land-Use Change: Theoretical and Modeling Approaches. The Web Book of Regional Science*, accessed January 17, 2014, http://www.rri.wvu.edu/WebBook/Briassoulis/chapter1%28introduction%29.htm#1.3.2.

Doygun, H. 2009. Effects of urban sprawl on agricultural land: A case study of Kahramanmaraş, Turkey. *Environmental Monitoring and Assessment, 158* (1–4), 471–478.

Dubovyk, O., Sliuzas, R., and Flacke, J. 2011. Spatio-temporal modelling of informal settlement development in Sancaktepe district, Istanbul, Turkey. *ISPRS Journal of Photogrammetry and Remote Sensing, 66* (2), 235–246. doi: http://dx.doi.org/10.1016/j.isprsjprs.2010.10.002.

Durand-Lasserve, A., and Selod, H. 2009. The formalization of urban land tenure in developing countries. In S. Lall, M. Freire, B. Yuen, R. Rajack and J.-J. Helluin (Eds.), *Urban Land Markets*, pp. 101–132. Dordrecht, The Netherlands: Springer.

FAO. 1995. *Planning for a Sustainable Use of Land Resources: Towards a New Approach*. Rome, Italy: FAO.

Farley, K. A., Ojeda-Revah, L., Atkinson, E. E., and Eaton-González, B. R. 2012. Changes in land use, land tenure, and landscape fragmentation in the Tijuana River Watershed following reform of the ejido sector. *Land Use Policy, 29* (1), 187–197. doi: http://dx.doi.org/10.1016/j.landusepol.2011.06.006.

Foley, J. A., DeFries, R., Asner, G. P., Barford, C., Bonan, G., Carpenter, S. R. et al. 2005. Global consequences of land use. *Science, 309* (5734), 570–574.

Gerlak, A. 2014. Policy interactions in human–landscape systems. *Environmental Management, 53* (1), 67–75. doi: 10.1007/s00267-013-0068-y.

Hersperger, A. M., and Burgi, M. 2010. How do policies shape landscapes? Landscape change and its political driving forces in the Limmat Valley, Switzerland 1930–2000. *Landscape Research, 35* (3), 259–279. doi: 10.1080/01426391003743561.

Hersperger, A. M., Gennaio, M. P., Verburg, P. H., and Burgi, M. 2010. Linking land change with driving forces and actors: Four conceptual models. *Ecology and Society, 15* (4).

Houghton, R. A., Hackler, J. L., and Cushman, R. M. 2001. *Carbon Flux to the Atmosphere from Land-Use Changes: 1850 to 1990*. Carbon Dioxide Information Analysis Center, Environmental Sciences Division. Oak Ridge, TN: Oak Ridge National Laboratory.

Li, L., Zhang, P.-Y., and Hou, W. 2005. Land use/cover change and driving forces in southern Liaoning Province since 1950s. *Chinese Geographical Science, 15* (2), 131–136.

Lu, D., Mausel, P., Brondizio, E., and Moran, E. 2004. Change detection techniques. *International Journal of Remote Sensing, 25* (12), 2365–2407. doi: 10.1080/0143116031000139863.

Mena, C. F., Bilsborrow, R. E., and McClain, M. E. 2006. Socioeconomic drivers of deforestation in the Northern Ecuadorian Amazon. *Environmental Management, 37* (6), 802–815. doi: 10.1007/s00267-003-0230-z.

Meyer, W. B., and Turner II, B. L. (Eds.). 1994. *Changes in Land Use and Land Cover: A Global Perspective*, p. 537, accessed January 23, 2014, http://books.google.nl/books/about/Changes_in_Land_Use_and_Land_Cover.html?id = X1pNRW6r0BoC&redir_esc = y.

Moser, S. C. 1996. A partial instructional module on global and regional land use /cover change: Assessing the data and searching for general relationships. *GeoJournal, 39* (3), 241–283.

Mottet, A., Ladet, S., Coqué, N., and Gibon, A. 2006. Agricultural land-use change and its drivers in mountain landscapes: A case study in the Pyrenees. *Agriculture, Ecosystems and Environment, 114* (2–4), 296–310. doi: http://dx.doi.org/10.1016/j.agee.2005.11.017.

Muttitanon, W., and Tripathi, N. K. 2005. Land use/land cover changes in the coastal zone of Ban Don Bay, Thailand, using Landsat 5 TM data. *International Journal of Remote Sensing, 26* (11), 2311–2323. doi: 10.1080/0143116051233132666.

Orenstein, D. E., Bradley, B. A., Albert, J., Mustard, J. F., and Hamburg, S. P. 2011. How much is built? Quantifying and interpreting patterns of built space from different data sources. *International Journal of Remote Sensing, 32* (9), 2621–2644. doi: 10.1080/01431161003713036.

Pielke, R. A., Marland, G., Betts, R. A., Chase, T. N., Eastman, J. L., Niles, J. O. et al. 2002. The influence of land-use change and landscape dynamics on the climate system: Relevance to climate-change policy beyond the radiative effect of greenhouse gases. *Philosophical Transactions of the Royal Society of London. Series A: Mathematical, Physical and Engineering Sciences, 360* (1797), 1705–1719.

Rurangwa, E. 2013. *Land Tenure Reform: The Case Study of Rwanda*. Paper presented at the Land Divided: Land and South African Society in 2013, in Comparative Perspective, Cape Town, South Africa.

Serneels, S., and Lambin, E. F. 2001. Proximate causes of land-use change in Narok District, Kenya: A spatial statistical model. *Agriculture, Ecosystems and Environment, 85* (1–3), 65–81. doi: http://dx.doi.org/10.1016/S0167-8809(01)00188-8.

Singh, A. 1989. Review article digital change detection techniques using remotely sensed data. *International Journal of Remote Sensing, 10* (6), 989–1003.

Sultana, S., and Powell, W. 2010. Land use. In B. Warf (Ed.), *Encyclopedia of Geography*, pp. 1727–1733. Thousand Oaks, CA: Sage.

Turner II, B. L., Meyer, W. B., and Skole, D. L. 1994. Global land-use/land-cover change: Towards an integrated study. *Integrating Earth System Science, Allen Press, 23* (1), 91–95.
Veldkamp, A., Lambinb, E. F. 2001. Predicting land-use change. *Agriculture, Ecosystems and Environment, 85*, 1–6.
Verburg, P. H., Schot, P. P., Dijst, M. J., and Veldkamp, A. 2004. Land use change modelling: Current practice and research priorities. *GeoJournal, 61* (4), 309–324.
Wannasai, N., and Shrestha, R. P. 2008. Role of land tenure security and farm household characteristics on land use change in the Prasae Watershed, Thailand. *Land Use Policy, 25* (2), 214–224. doi: 10.1016/j.landusepol.2007.07.003.
Williamson, I., Enemark, S., Wallace, J., and Rajabifard, A. 2010. *Land Administration for Sustainable Development*. Redlands, CA: ESRI.
Zevenbergen, J. 2002. *Systems of Land Registration: Aspects and Effects.* PhD thesis. Delft, The Netherlands.

13

Environmental Protection via Land Administration

Libia Y. Romero Lara, Jaap Zevenbergen, and Kees Bronsveld

CONTENTS

Introduction

Protection of the environment on private lands, and lands where people hold private interests, embodies both a challenge and an opportunity for any government (Shogren et al. 2003). The challenge is the complex task of understanding and integrating the arrangement of diverse rights including "property rights, access rights, harvesting rights, management rights, exclusion rights and alienation rights" (Sandberg 2007), among others. The opportunity is the improved level of care afforded to the environment that private rights to land and resources can provide (United Republic of Tanzania 1997).

Achieving desired results in the implementation of environmental policies is considered to require collaboration between private right holders and other actors with vested interests in the land. Whenever the points of agreement and disagreement between these actors have not been considered and identified, there is a high chance that accomplishment of the policies will be frustrated (Cocklin et al. 2007). For this reason, it is considered important to improve knowledge bases about the diverse range of rights held in the areas of interest, while also promoting inclusion of the right holders in participatory decision-making processes. In this respect, Williamson (2001) identifies the crucial role that cadastres can play in understanding and administering the relationships between people and land.

However, while such collaborative approaches are promoted, it is still generally not clear to what extent existing rights, interests, and uses over land determine the performance of environmental policy implementations (Wanitzek and Sippel 1998). Prescribed environmental measures, and the stakeholders involved, often focus merely on environmental implications and underestimate the importance of other critical variables: the underlying role of land rights, establishment of quantifiable indicators regarding community and individual rights, as well as implementation of instruments designed to deal with those rights (Balint 2006) are often neglected. Therefore, it is necessary to improve understandings of the relationship between land rights, land uses, and environmental measures—potentially using alternative analysis tools relating to land, and not merely the cadastre.

The aforementioned sentiments are especially relevant in a country like Tanzania, where the presence and strength of various types of formal and traditional rights are diverse (Wanitzek and Sippel 1998). Tanzania is vastly dependent on its natural resources, which play an important role in terms of social and economic goods and services in its national economy (United Republic of Tanzania 2009). Disputes regarding land in or around protected areas in Tanzania are well known. This is a consequence of overlapping interests between local communities throughout the Tanzanian countryside and government institutions that manage and administer these protected areas (Wanitzek and Sippel 1998). The need to deal with possible conflicts becomes evident when one considers the enormous space covered by protected areas in Tanzania: 39.6% of the total land area is protected (World Resource Institute 2009).

This chapter aims to challenge conventional understandings of the conflicting relationship between environmental measures and private rights and additionally propose new mechanisms for dealing with it. For this, results from a case study in a protected area in Tanzania, Saadani National Park (SANAPA), are used. The data collected from the study, plus the literature accompanying review, are interpreted using an adaptation of the DPSIR (driving forces–pressures–state–impacts–responses) framework. This method assumes cause–effect relationships between interacting components of social, economic, and environmental systems (UNEP-GRID Arendal 2009). Following the identification of the elements assumed to be influential in the effect relationships between the establishment of SANAPA and the rights of local communities, some indicators are drawn to enable predictions of conflicts. This chapter is also innovative in that it applies the Dynamic Actor Network Analysis (DANA) software, a tool for supporting the analysis of actors involved in a process through the modeling of their perceptions. Using the aforementioned approaches, it was considered possible to identify suitable instruments for addressing conflicts between environmental measures and private land interests.

Overall, the empirical data collected allowed the following: (1) the identification of elements that have more influence in the establishment of an environmental measure in a protected area containing existing rights,

(2) establishment of possible indicators for conflicts between application of environmental measures and existing land rights, (3) formation of understandings of the way in which stakeholders' perceptions influence the implementation of an environmental measure in a protected area, and (4) proposal of possible instruments that could be included in the design of an environmental measure design.

Theoretical Perspective

Although environmental protection is a subject that generates much positive sentiment globally, it involves significant debate due to the restrictions it places on human activities. These involve limiting different types of rights and uses over natural resources. They stimulate resistance when either the economic productivity of land or the established way of life is disturbed (Doremus 2003).

Among the range of available alternatives for environmental protection, the establishment of protected areas—such as national parks (Cernea and Schmidt-Soltau 2006) and wildlife reserves—is a largely accepted approach by national and international organizations (Udaya Sekhar 2003). The institution of national parks has facilitated the task of prevention of loss of biodiversity and wildlife destruction caused by development and land conversion. However, because of the level of enforcement, right holders—particularly in developing countries—are susceptible to the establishment of such areas (Skonhoft 1998). They not only lose their access to natural resources, resulting most of the time in forced livelihood changes, but also are exposed to forced displacement:

> compulsory removal initiated when a project's need for 'right of way' is deemed to override the 'right to stay' of the inhabiting populations (Cernea and Schmidt-Soltau 2006).

Factors like landlessness, joblessness, homelessness, marginalization, food insecurity, loss of access to common property, and social disarticulation are commonly associated with population displacements after the establishment of a protected area (Cernea and Schmidt-Soltau 2006).

Several studies (Langholz and Lassoie 2001; Lindsey et al. 2005; Songorwa 1999) illustrate how the acceptance of protected areas by local people depends on harmonizing their own interest with the goals of the nature reserve.

The acquisition of land rights is a strategy frequently adapted by governments for the establishment of protected areas such as parks. According to the situation in place, the way it is implemented may vary from a voluntary basis, through donation or purchase at a mutually agreed price, to condemnation (Doremus 2003).

There exists no unique formula for selecting suitable approaches to reach the multifunctional goals that environmental protection seeks. The context, including landscape, land tenures, and governance arrangements constitute some of the factors that will lead to the selection of a particular strategy (Cubbage et al. 2007). However, whatever conservation approach is taken rights holders—and in general those who have been disturbed with the establishment of the protected areas—should be informed, educated, and taken into account (Van Gossum et al. 2005). Sensitivity to the specific conservation goals and the local context and continuous monitoring are also key issues (Wells et al. 1992).

In Tanzania, all land is considered as public land with value and kept by the president for the general public. Since 1969, and under the Government Leasehold Act, "Rights of Occupancy" have been issued with development conditions enforced. Customary rights of occupancy are also recognized. Under the National Land Policy (NLP), revocation of rights of occupancy might occur in cases of public interest, and it should include compensation on the bases of cost-opportunity. The second edition of the NLP includes the establishment of means for protection of sensitive areas such as national parks, and it states that such areas are not the subject of allocation to individuals. Unfortunately, the registration of statutory allocation of these areas does not take place, causing, in most of the cases, encroachments and alienations.

As soon as an area is declared a national park, any previous claims on the land and all existing rights are vested in the president (Tanzania National Parks, and Department of Planning and Projects Development 2003). The NLP has recognized weaknesses in provisions on compensation leading to complaints about rates, delays in payment, and the nonemployment of alternative assessment techniques.

Generally speaking, Tanzanians highly approved the establishment of protected areas: they considered them part of the country's national heritage (Wanitzek and Sippel 1998). However, as previously discussed, whether this was also at the exclusion of the preexisting livelihoods in those areas, or neighboring areas, is a more debated matter.

Methodology

A case study was considered the most relevant research strategy to utilize. The fieldwork was conducted in Tanzania, in the city of Dar es Salaam and the small villages of Uvinje and Buyuni, inside the park, and Saadani village, placed on the border zone of the park. The main criteria for selection of the study area were (1) existence of high levels of enforcement of environmental measures in the area of study and (2) presence of conflicts or observable effects on local people and/or their rights.

Data were collected from interviews, observations, and secondary sources. Data were collated and analyzed by first representing the different components of the SANAPA right holders' system and their interactions using a modified DPSIR analysis. DPSIR uses processes to explain the interrelations between human activities and environment (Nilsson et al. 2009). Five base categories, logically bonded, are prescribed: driving forces, pressure, state, impact, and responses (European Environment Agency 1999). The framework in this context needed to be adapted: although DPSIR is an established and frequently employed framework for understanding the roots and scope of environmental problems (Niemeijer and de Groot 2008), it is usually concentrated merely on environmental issues, with little attention given to social and economic issues (Svarstad et al. 2008). Considering that the changes were done only in the subject represented, and that all the aforementioned three issues are relevant for this study, the essence of the framework is preserved; thus, it is assumed to remain reliable. As a matter of fact, since its very beginnings in 1979 the conceptual framework known today as DPSIR has had different adaptations to make it suitable for different contexts, as described by Gabrielsen and Bosch (2003).

Once all the data collected were structured under the different categories that DPSIR prescribes, identification of the variable indicators was undertaken. The development of indicators is based on the modeling of stakeholders' perceptions, based on an actor-oriented approach. The indicators were essentially quantitative, but also qualitative, in the case of those characteristics relevant to the study with nonquantifying qualities. The assumption behind this approach is that once the perceptions and views of different actors or stakeholders involved in a policy issue are exposed (in a model in this case) useful understandings are gained for identifying the core issues that develop into conflicts. Consequently, responses that might reduce the level of conflicts can be identified. However, once the analysis is based on perceptions it is quite hard to talk about "objective reality" (Bots et al. 1999); therefore, the conception of the model presented is fully based on inferences about the data provided by the interviewees.

The selected tool to model stakeholders' perceptions was DANA, software developed at the Delft University of Technology, Netherlands, by Pieter W.G. Bots. In DANA, "causal relations diagrams" are used to model the perceptions of stakeholders; those diagrams are no more than factors and mechanisms relevant to each actor and the causal relations between them (Hermans 2004). In DANA, actors represent different people, organizations, or general groupings that play a role in the issue being analyzed. Two groups of actors can be modeled using this tool: stakeholders and agents. Changes introduced by actors' actions, or external influences, can take place in the issue analyzed. When these changes affect an actor's interest, such an actor is called a stakeholder; otherwise, such an actor is an agent (Bots 2009).

The stakeholders included in the analysis were Tanzania National Parks (TANAPA) and local villagers. They both influenced and are influenced by the system. Due to the diversity of circumstances and effects caused, local villagers were divided into four categories for a deeper model and analysis: (1) villagers from Saadani, (2) villagers living inside the park, (3) villagers who resettled, and (4) villagers who moved to Saadani village looking for opportunities brought by the park. The perceptions of each stakeholder needed to be considered taking into account the limited data available and the people interviewed. It is assumed that the perceptions taken from the interviews represent the perceptions of a certain actor type in the model.

The final step of the analysis was to propose possible instruments to reduce the level of conflict between stakeholders. The feasibility of application of those instruments was not part of the scope of this research.

Results

Results from the DPSIR conceptual framework analysis are first discussed. As per the framework, the following categories are used: driving forces, pressures, state, impact, and responses.

The first elements that came into place as driving forces were the policies adopted by the government of Tanzania regarding environmental issues, specifically those dealing with national parks and land. These policies were the basis for the current regulating act that subsequently allowed the creation and administration of the network of protected areas, already in place in Tanzania, with emphasis on disapproval of human settlement inside them. Population factors such as increase, irregular distribution and vicinity to protected areas (attracted by the potential benefits in tourism) have also to be considered as Driving Forces. The devotion of rural communities to exploitation of natural resources can also explain their tendency to settle near protected areas.

Meanwhile, there are two pressures that represent the way drivers affect the system: (1) establishment of SANAPA and (2) presence of human settlements. In 1969, and after a demand from the villagers themselves, the Saadani ecosystem was declared as a game reserve in an attempt to protect the wildlife, which was in serious threat due to permanent hunting. People were allowed to keep on using natural resources, without farming or settlement. However, later in 1998 TANAPA revealed its intention of increasing the area and level of protection of the Saadani ecosystem. The Saadani ecosystem is home to approximately 35,000 people mostly distributed in 10 main settlements, spread out around the park (Tanzania National Parks, and Department of Planning and Projects Development 2003). According to the opinions collected from different villagers, local communities entirely rely on the use of natural resources from the region, which introduces a strong pressure on the system, leading to a threat for the protection and conservation of the entire ecosystem and then to conflicts with the community that used to have free access to those resources. Roettcher (2001) cited by Ally Hassan (2005) already refers to the situation, illustrating the dependence of local communities on the protected area to obtain firewood, water, building materials, and dwarf palm.

The state or current condition found can be summarized in three items: (1) uncertainty about boundaries, (2) resettlements, and (3) increasing conflicts between SANAPA administration and local communities. People in Saadani do not know where they can keep their livestock, where they can grow crops, or where they can go and collect natural resources. None of the villagers interviewed have ever been told about the boundary of the park, nor have they even seen a map of it. They simply do not know where the boundary of their village is. As already discussed, eviction of all those living inside a national park is the next step once an area is declared. In the case of SANAPA, that was not different. Some conflicts also arise due to crop damage by wild animals; illegal use of natural resources; and people seeking water, firewood, and other resources to build their houses and for other general purposes.

The impact of this situation is reflected in the increased limitation of livelihoods and a negative perception about the park. The economy of the area around the Saadani ecosystem is quite limited: fishing, limited cropping, and small livestock keeping are the available options. After gradual limitations in uses in the park (from open access to game reserve, and then national park) in terms of access to natural resources, their main "supplier" has been restricted. Therefore, their livelihoods have been limited. People who settled around Saadani claim that they have been living with natural resources from the Saadani ecosystem for a very long time; therefore, they already have "rights" over them. Another issue to consider is the low generation of income through the development of tourism in the park. Compared to other national parks in Tanzania, like Serengeti or Kilimanjaro, the tourism rate in SANAPA is quite low, mainly because of poor and unreliable access to the area and the limited quality of the beach. Moreover, according to the available information for this study SANAPA administration has provided no mechanisms to guarantee that some of the revenue generated is directed to community development projects. In general, SANAPA administration is not highly rated among most of the villagers, both outside and inside the park. The lack of consultation and community involvement in decision-making processes has caused people to distrust the administration.

Finally, given the state and its impact on the system, some actions were taken or are planned to be taken. Responses like the implementation of community-based conservation address many of the categories in the DPSIR chain.

The identification of elements in each category of the DPSIR conceptual framework allowed the selection and definition of indicators (Ojeda-Martínez et al. 2009); Table 13.1 contains the list of indicators that correspond to the elements analyzed.

Regarding the results from the modeling and analysis of perceptions, as discussed, this can be seen as a good way of identifying divergences in the

TABLE 13.1

Indicators according to DPSIR Framework of SANAPA and Land Rights Holders

Indicator	Description
Driving Forces	
Population growth	Estimated percentage of the rate of population growth per year
Population density	Number of people settled around the protected area per square kilometer
Economic activities	Economic activities taking place around the protected area and that involve the use of natural resources
Economic instruments	Type of economic instruments provided in the environmental policy
Type of rights	Type of rights (customary/statutory) and its characteristics
Pressure	
Category of conservation	Type of category of the protected area
Size of protected area	Area in square kilometers of the protected area
External population	Number of people neighboring to the protected area
Internal population	Number of people living inside the protected area
Minimum distance of settlements	Distance in kilometers from the protected area to the closest settlement
State	
Clarity about boundaries	Percentage of the population that recognizes the boundaries of the protected area
	Existence of conflicting maps or boundary descriptions
Participatory decision-making process	Percentage of the population consulted about the decisions to make
Representativeness	Percentage of each sector or group of the community that took part in the decision process
Socialization	Means used to inform the community about the decision and projects to carry out
Compensation fairness	Rate of the compensation paid against the commercial value of each possession
Compensation satisfaction	Percentage of the population that considers the compensation paid was fair.
Buffer zones	Area in square kilometers of the buffer zones around the protected area
Crop damage	Number of incidents reported regarding crop damaged by animals.
Land availability	Extension of land in kilometers available for normal community development.
Impact	
Restriction of resources	Type of restriction imposed on local communities regarding the use of natural resources
Training in alternative economic activities	Number of projects developed
	Percentage of local people involved
	Budget invested

(Continued)

TABLE 13.1 (*Continued*)

Indicators according to DPSIR Framework of SANAPA and Land Rights Holders

Indicator	Description
Capacitating about wildlife–human coexistence	Number of projects developed
	Percentage of local community involved
	Budget invested
Responses	
Tourism benefits	Percentage of the revenues obtained from tourism allocated to projects in benefit of the community
Job generation	Number of local people hired for administration of the protected area
Community development	Status of the main social facilities of the community (school, hospital, etc.)
	Budget in project to improve the status of the social facilities of the community
Infrastructure development	Status of the infrastructure
	Budget in project to improve the status of the infrastructure

views of different stakeholders related to an issue, allowing in turn the recognition of conflict between them. Five stakeholders were identified in the "system SANAPA-local people."

The first stakeholder identified was TANAPA. It is the organization that manages and regulates the national parks in Tanzania. The perception graph was derived from the literature. Four goals were identified: (1) maintenance of flora and fauna of SANAPA safe from conflicting interests of a growing population, (2) community development, (3) preservation of natural heritage, and (4) reduction of conflicts. All these goals are deeply influenced by the management of SANAPA and by interaction with local communities.

The second stakeholder group identified was the group of people who were living in Saadani, even prior to the establishment of the previous game reserve. They report having faced a gradual limitation in their rights and the area of their village. They recognize their illegal access to the park as the only way to access land and the resources needed for life, once their traditional rights were reduced. They support the preservation of the natural heritage and want their incomes and community development programs to be increased. Land tenure insecurity should be reduced, and their traditional rights and livelihoods should not be lost.

The third stakeholder group comprises those people who are living, and continue living, inside the park after its establishment, despite what is envisaged in the legislation. In general, their goals coincided with those of the previous actor; however, they also do not want to see more changes in rights to land and resources.

The fourth stakeholder group is the group of villagers who, contrary to the second category, were told to move out of SANAPA after receiving compensation. They appear to be the most affected villagers neighboring SANAPA. They were living in an area declared as protected, but unlike the previous actor they were not allowed to stay inside the park and were forced to settle somewhere else. Some of them, trying to be as close as possible to their original place of residence, settled in Saadani village. As expected, they do not want their livelihoods and access to land and land resources to be reduced, but also they seek to do away with resettlements and impoverishment.

The fifth stakeholder group, the last stakeholder analyzed in this study, is the group of villagers who spontaneously migrated from elsewhere and went to Saadani after they learned about the establishment of the park, searching for opportunities that tourism would bring. The establishment of the park is believed to increase the investment in community projects. Thus, villagers expect an improvement in their quality of life. They are mostly interested in the opportunities from the park itself to improve their economic condition and increase their incomes.

Meanwhile, in addition to the five stakeholder groups some other agents were identified, but they were not included in the analysis. In summary, these included (1) the president of Tanzania, who is responsible for the declaration of any protected area, and he or she can also modify the boundaries of a national park; (2) Ministry of Tourism and Natural Resources (MTNR), which heads TANAPA; (3) Tanzanian Wildlife Division; and (4) district local government. All nine actors are linked together, creating a network as seen in Figure 13.1.

After modeling five of the nine actors' perceptions, two analysis cases were conducted: (1) TANAPA and integrated local villagers' perceptions, meaning that all the four categories of local people were grouped in the same perception view; and (2) TANAPA and different villagers' perceptions, in which a simplified view of each different villager's category was considered. The first

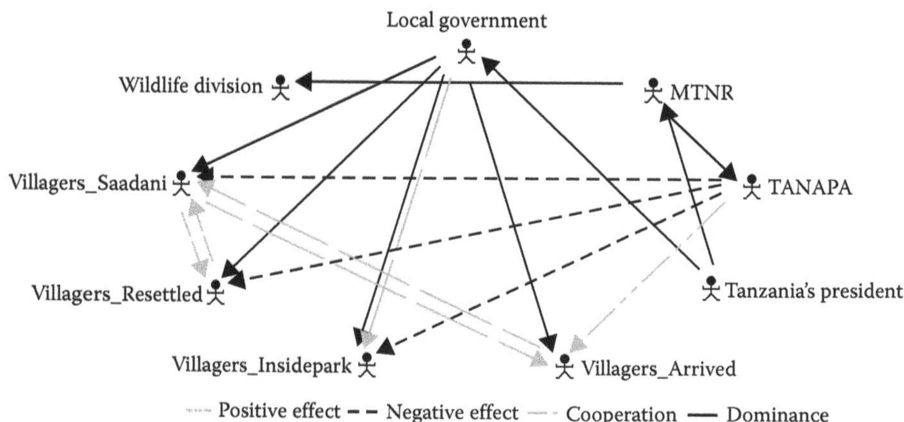

FIGURE 13.1
Actor relations in the establishment of Saadani National Park.

analysis case revealed that the stakeholders diverge only in the actions that should be taken, not in their goals.

In summary, local people experienced dissatisfaction with the extinction of their traditional rights with the establishment of SANAPA, whereas TANAPA experiences dissatisfaction with the illegal access of local people to the park. According to the analysis in DANA, the inferred best strategy from the point of view of local people is to increase access to the park and have more investment in community projects by TANAPA. Their worst strategy would be to keep things as they are. On the contrary, from the point of view of TANAPA the best strategy would be to reduce the illegal access to the park of local people, while preserving their investment in community projects. This clearly shows the divergence in relation to access to the park according to each stakeholder: local people want more of it, whereas TANAPA wants to reduce it. The differences in experience of one stakeholder when the best strategy of the other has taken place allowed identifying that apparently TANAPA is more tolerant to the illegal access to the park by local people than the latter to the extinction of their traditional rights.

The second analysis case (Figure 13.2) reiterated the findings of the first, with identified conflicts in the actions to be taken.

An interesting result of this analysis was that there is not only conflict between TANAPA and the different villagers but also conflict between villages. This means that different categories of villagers also diverge in the

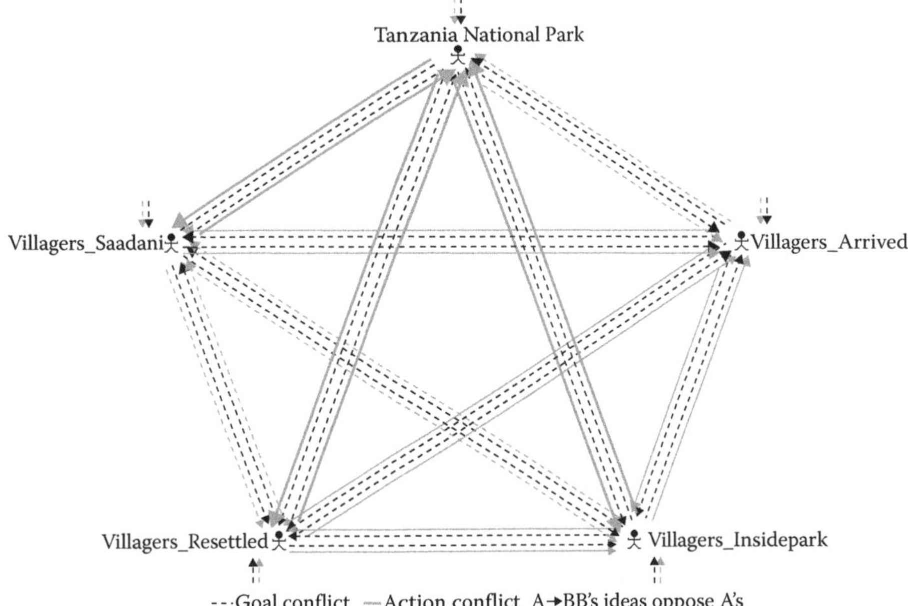

FIGURE 13.2
Goal and action conflict relation between actors—analysis 2.

actions that should be taken, and in what they experience from these actions. The category of villagers that mostly rejects the actions to be taken for TANAPA comprises those who stayed inside the park, which can be explained by the "feeling" they have that they might experience an eviction at any time.

Comparing the results of both analyses, it was noticed that the first case showed an expected level of conflict between TANAPA and local communities and the analysis of the second case allowed identification that actually the different categories of local communities differ in the way they perceive the issue of the establishment of the park. Thus, the level of disagreement between them and the administration of the park, regarding the way things should be done, also differs. This is an expected result, taking into account that each group of villagers was not affected in the same way.

Finally, regarding the results from the process of selecting policy instruments, it is apparent that a variety of instruments could be adopted according to the type of policy and measures implemented. The selection of the optimum instrument for each case is an important task. It should be the most efficient, effective, equitable, and acceptable instrument for both communities and authorities (Australian Public Service Commission 2009). A list of possible instruments to be applied is proposed (Table 13.2). These are

TABLE 13.2

Proposed Instruments to Be Included in the Implementation of SANAPA

Instrument	Description	Conflict Addressed
Using Markets		
Targeted subsidies	Subsidies for the development of alternative economic activities that do not depend on natural resources inside the park.	Illegal access to SANAPA. Reduction of livelihoods.
Creating Markets		
Water easement	This is a relevant instrument considering that rivers running inside the park are the only access that the local communities have to freshwater.	Illegal access to SANAPA. Extinction of traditional rights of local communities. Reduction of livelihoods.
Environmental Regulation		
Zoning	Establishment of a transition zone around the villages surrounding the park where access to certain natural resources is allowed with certain restrictions.	Illegal access to SANAPA. The extinction of traditional rights of local communities.
Engaging the Public		
Information provision	Continuous diffusion of printed materials about the relevance of SANAPA and its management.	The lack of proper communication between the parties involved.
Community participation	Periodic multi-way dialogues between TANAPA staff and local communities.	

compiled from a synthesis of the results from the DPSIR framework analysis, DANA modeling process, and the four categories of instruments proposed by the World Bank (1997) and cited by Sterner (2003) and SIDA (2003).

Discussion

Before the establishment of SANAPA in 2005, there were already various communities settled in the area with a long tradition of use and access to land and natural resources. These land uses and access regimes are now illegal. Moreover, villagers have little or no clarity about the boundaries of the park and they do not have any security with respect to the rights they actually hold. Although the Wildlife Policy calls for survey and acquisition of land title deeds after the establishment of protected areas, no registration was found in the case of SANAPA. This fact might be explained by two issues. One is that most of the rights that people had were customary, and not registered: when the park was established, having no preceding registration, there was no registration to update. Another issue might be the lack of communication between environmental and land agencies.

The results of the analysis performed with the DPSIR framework and DANA software coincide with the discussion about possible conflicts emerging in the event of the establishment of protected areas, as raised by Vatn (2005) and Bergseng and Vatn (2009). Contrasting with the classification of conflicts that these authors proposed with the results obtained in this research, it can be suggested that TANAPA and local communities experience four different disagreements: (1) conflicts of interests regarding the compensation paid, (2) conflicts of interests regarding the extinguishment of rights, (3) conflicts of facts regarding the decrease of livelihoods, and (4) conflicts of value regarding their perceptions.

The conceptual framework implemented in this research proved to be a very useful tool to capture the main issues associated with the subject of study. All the elements were presented in a very simple but highly illustrative manner. This is the major advantage of DPSIR, together with its flexibility. Nonetheless, as discussed by Svarstad et al. (2008), the major disadvantage of DPSIR can also be its weakness: it may lead to simplistic representations of complex issues as the one dealt with in this research.

The integration of the DPSIR framework with the modeling of perceptions is conceived as a complementary approach to increase the analysis and understanding of this complex tenure problem. While DPSIR provides a generalized vision, the modeling of perception graphs provides deep perspectives on the actors involved. Both approaches, either individually or combined, can be applied in the field of land administration. Among the advantages, one could better represent complex land tenure situations (or the existing continuum of land rights), seek to value security of tenure, better understand the context in which a cadastre and a land register operate, and in general analyze the effects of any land reform (Deininger 2003).

The formulation of indicators from the framework coincides with the approach supported by Balint (2006). He highlighted the importance of identifying relevant factors to improve community-conservation approaches close to protected areas and used them in the form of indicators to foresee the success or failure of initiatives taken for environmental conservation or in the context of land administration: land management initiatives. The indicators in this research observe some characteristics that were discussed by Veleva et al. (2001) related to quality: they were abstracted based on the data collected in the field, which despite the encountered limitations are assumed to be accurate. They were kept simple, and easy to implement and evaluate.

Meanwhile, including the perceptions of the most relevant actors involved in the system provided insights on the cause–effect relations between actions and the conflicts between them, with respect to SANAPA. This was the main reason for the use of this approach: such valuable output would have stayed hidden with a different approach.

Given the type of the research carried out, the analysis and interpretation given to the data collected included a number of assumptions. The perception graphs do not reflect the whole view and opinions of the actors analyzed. This is due to limitations in data collection and the inability of the method to extract the complexity embedded in the issue being studied and to adequately model how people perceive it. For example, the perception graphs in DANA were simplified to enable calculations to be performed. Although this may have resulted in a reduction of the reliability of the results, the complexity of interpreting the results was also reduced, which resulted in a more comprehensible understanding.

In general, the results of the analysis performed by DANA cannot be taken as complete, although they provide a good basis for understanding the issue through discussions of the outputs. Furthermore, the validation of the model should ideally be performed by the actors themselves, which would give relevant inputs to improve an initial approach; also, the analysis should include the perspectives of different analysts. Discussions with the actors represented are the best way to ensure the quality of the results. The analysis of two simplified models, one assuming all villagers as one actor and the other assuming an independent view of the identified different types of villagers, is considered as a way of validating the models constructed.

Policy instruments as tools that assist the achievement of policy objectives are also worthy of discussion, particularly their relevance and impact on the field of land administration. Due to the complexity of the issue analyzed, the proposed instruments supposed a combination of different policy responses, each addressing detected conflicts concerning the establishment of SANAPA. This approach is assumed to work better than trying to tackle each problem individually. However, the identification of the best instruments demands a judicious study not only about the impact the instruments will have but also about the feasibility in cost, administration, and time, among other factors. Due to time constraints, such a study was not carried out before proposing the instruments and neither was there a joint discussion with the actors potentially involved in the implementation. This, of course, undermines their validity and approval. Nevertheless, they are considered a good starting point for debate of the parties involved. Using the output of the DPSIR framework as a background, and the output of DANA analysis as the guider, it was possible to provide a more reliable, although not complete, group of instruments.

The topic of this research is a complex issue that can be addressed in several ways and with different degrees of extension and depth. The approach adopted was extensive enough to go from explanation of constitutive elements and identification of conflicts through to proposal of instruments to reduce those conflicts. It was considered deep enough to allow the use of a conceptual framework to explain the relationship of such elements, and to represent the views and perceptions of the actors involved. However, the methodology adopted and results produced could be improved in precision, with a wider and more detailed information capture exercise, especially in relation to the modeling of actors' perceptions—a technique that imposes a somewhat higher degree of subjectivity than a conceptual framework. An influential element for the reliability of these methods, apart from the data collected, is the follow-up discussions with the informants in the case of DPSIR and with the actors in the case of DANA. The same applies to the policy instruments.

In general, the two main methodologies applied after fieldwork suited the intention of responding to the research questions and seemed to complement each other. They also gave a basis to research more about them and expand their potential application in the land administration domain. Inclusion of spatial analysis could also add more valuable and interesting outputs.

Conclusion

Using an adaptation of the DPSIR conceptual framework allowed identification and explanation of the elements that, according to the data gathered, appear influential in defining the effects of a specific environmental measure on the existing rights to land of a local community. The elements were listed, on a causal basis, under five different categories: driving forces, pressures, state, impact, and responses. Even though data limitations and assumptions made during the construction of the framework may decrease the reliability of the output, the framework seems to capture the relevant elements associated with the subject of study.

Depictions and analyses of perception graphs of the most relevant stakeholders were carried out using the DANA software. The graphs were developed using the data collected, the analyst's view, and several assumptions regarding how much one factor influenced the situation. Though validation of the models by the included stakeholders was not feasible, the outputs of the analysis appear to be consistent in reporting a conflict of actions in the two models constructed. The use of DANA imposed limitations on the complexity of the models and of the actor's perception graphs. They should be kept simple to not affect software performance, and also because the results of the analysis are more difficult to interpret.

After the analysis of perception graphs and the identification of elements creating tension between stakeholders, several policy instruments—ones considered to most likely address the identified conflicts—were proposed as responses to reduce the conflicts.

In general, the DPISR framework and DANA modeling methodologies seem to be complementary. The first allows analysis of the general elements relevant to a policy subject, whereas the second helps to place those elements in the way each actor perceives them. Due to their being based on causal relations, both DPSIR framework and DANA software need more data to give more reliable outputs. The simultaneous use of different methodologies allows cross-checking and hence improves validity and reliability of results.

Overall, several recommendations are derived from the research. First, to improve the quality of outputs for similar future research the fieldwork should ideally be carried out in two steps. An initial step should aim to collect and identify the main elements and their relations in the system being studied. Interviews should be flexible and incorporate brainstorming. An initial DPSIR framework and actor's perception model should be drawn. These drafts should be the basis for a second round of data collection. Second, although DPSIR has proved to be a flexible and adaptable framework, the adaptation of concepts should be judiciously analyzed to determine the degree of distorted results it might have, and thus the implications for quality outputs. Third, after the identification of indicators to understand the effect of an environmental measure on the rights of local communities, it is recommended to identify the desired or ideal state for each of the elements that are included in the analysis. Thus, comparative studies can be performed between the ideal and the current condition of the elements and goals. Weighting of each indicator could also be undertaken. Fourth, due to the fact that actors behave according to their interests and values, the modeling of actor perceptions should first identify the different types of interests. Those who share values and interests can be jointly modeled and represented. Fifth, further analysis of the acceptability and efficacy of the proposed instruments might be included in the models of perceptions. Sixth, additional study is needed on the reliability of the results of the analysis of actors' perception graphs that DANA generates. Finally, the applicability of DANA software, and in general of the concept of perception modeling, to study the aptitudes and viewpoints of stakeholders regarding land administration issues (e.g., registration, land titling, and land consolidation projects) could be further trialed.

References

Ally Hassan, N. 2005. Experiences of land use conflict management strategies in rural Tanzania: The case of protection and utilization of Saadani ecosystem. University of Dar es Salaam–University of Dortmund.

Australian Public Service Commission. 2009. *Smarter Policy: Choosing Policy Instruments and Working with Others to Influence Behaviour*. Canberra, Australia: Australian Public Service Commission.

Balint, P. 2006. "Improving community-based conservation near protected areas: The importance of development variables." *Environmental Management* 38 (1):137–148.

Bergseng, E., and A. Vatn. 2009. "Why protection of biodiversity creates conflict—some evidence from the Nordic countries." *Journal of Forest Economics* 15 (3):147–165.

Bots, P., M. van Twist, and J.H. Ron van Duin. 1999. Designing a Power Tool for Policy Analysts: Dynamic Actor Network Analysis. Paper read at Proceedings of the 32nd Hawaii International Conference on System Sciences, Maui, HI.

Bots, P.W.G. 2009. DANA: Dynamic Actor Network Analysis, 2009 2004, accessed September, 2009, http://dana.actoranalysis.com/.

Cernea, M.M., and K. Schmidt-Soltau. 2006. "Poverty risks and national parks: Policy issues in conservation and resettlement." *World Development* 34 (10):1808–1830.

Cocklin, C., N. Mautner, and J. Dibden. 2007. "Public policy, private landholders: Perspectives on policy mechanisms for sustainable land management." *Journal of Environmental Management* 85 (4):986–998.

Cubbage, F., P. Harou, and E. Sills. 2007. "Policy instruments to enhance multi-functional forest management." *Forest Policy and Economics* 9 (7):833–851.

Deininger, Klaus. 2003. Land policies for growth and poverty reduction. A World Bank policy research report. Washington, DC: World Bank Group. http://documents.worldbank.org/curated/en/2003/06/2457830/land-policies-growth-poverty-reduction.

Doremus, H. 2003. "A policy portfolio approach to biodiversity protection on private lands." *Environmental Science and Policy* 6 (3):217–232.

European Environment Agency. 1999. *Environmental Indicators: Typology and Overview*. Copenhagen, Denmark.

Gabrielsen, P., and P. Bosch. 2003. Environmental Indicators: Typology and Use in Reporting. EEA internal working paper. Copenhagen, Denmark: European Environment Agency.

Hermans, L.M. 2004. "Dynamic actor network analysis for diffuse pollution in the province of North-Holland." *Water Science and Technology* 49 (3):205–212.

Langholz, J., and J. Lassoie. 2001. "Combining conservation and development on private lands: Lessons from Costa Rica." *Environment, Development and Sustainability* 3 (4):309–322.

Lindsey, P.A., J.T. du Toit, and M.G.L. Mills. 2005. "Attitudes of ranchers towards African wild dogs *Lycaon pictus*: Conservation implications on private land." *Biological Conservation* 125 (1):113–121.

Niemeijer, D., and R. de Groot. 2008. "Framing environmental indicators: Moving from causal chains to causal networks." *Environment, Development and Sustainability* 10 (1):89–106.

Nilsson, M., H. Wiklund, G. Finnveden, D.K. Jonsson, K. Lundberg, S. Tyskeng et al. 2009. "Analytical framework and tool kit for SEA follow-up." *Environmental Impact Assessment Review* 29 (3):186–199.

Ojeda-Martínez, C., F.G. Casalduero, J.T. Bayle-Sempere, C.B. Cebrián, C. Valle, J.L. Sanchez-Lizaso et al. 2009. "A conceptual framework for the integral management of marine protected areas." *Ocean and Coastal Management* 52 (2):89–101.

Sandberg, A. 2007. "Property rights and ecosystem properties." *Land Use Policy* 24 (4):613–623.

Shogren, J.F., G.M. Parkhurst, and C. Settle. 2003. "Integrating economics and ecology to protect nature on private lands: models, methods, and mindsets." *Environmental Science and Policy* 6 (3):233–242.

SIDA. 2003. Instruments for Environmental Policy. Environment Policy Division—Swedish International Development Cooperation Agency.

Skonhoft, A. 1998. "Resource utilization, property rights and welfare—wildlife and the local people." *Ecological Economics* 26 (1):67–80.

Songorwa, A.N. 1999. "Community-based wildlife management (CWM) in Tanzania: Are the communities interested?" *World Development* 27 (12):2061–2079.

Sterner, T. 2003. *Policy Instruments for Environmental and Natural Resource Management*. Gothenburg, Sweden: RFF Press.

Svarstad, H., L.K. Petersen, D. Rothman, H. Siepel, and F. Wätzold. 2008. "Discursive biases of the environmental research framework DPSIR." *Land Use Policy* 25 (1):116–125.

Tanzania National Parks, and Department of Planning and Projects Development. 2003. Saadani National Park Management Zone Plan Environmental Impact Assessment. Arusha, Tanzania.

UNEP-GRID Arendal. 2009. Cities environment report on the Internet, November 25, 2003, accessed August 17, 2009, http://www.ceroi.net/.

Udaya Sekhar, N. 2003. "Local people's attitudes towards conservation and wildlife tourism around Sariska Tiger Reserve, India." *Journal of Environmental Management* 69 (4):339–347.

United Republic of Tanzania. 1997. National Environmental Policy. Dar es Salaam, Tanzania.: Vice President's Office.

United Republic of Tanzania. 2009. National Website of the United Republic of Tanzania 2007, accessed June 21, 2009, www.tanzania.go.tz.

Van Gossum, P., S. Luyssaert, I. Serbruyns, and F. Mortier. 2005. "Forest groups as support to private forest owners in developing close-to-nature management." *Forest Policy and Economics* 7 (4):589–601.

Vatn, A. 2005. *Institutions and the Environment*. Cheltenham, United Kingdom: Edward Elgar.

Veleva, V., and M. Ellenbecker. 2001. "Indicators of sustainable production: Framework and methodology." *Journal of Cleaner Production* 9 (6):519–549.

Wanitzek, U., and H. Sippel. 1998. "Land rights in conservation areas in Tanzania." *GeoJournal* 46 (2):113–128.

Wells, M., K. Brandon, and L. Hannah. 1992. *People and Parks: Linking Protected Area Management with Local Communities*. Washington, DC: World Bank.

Williamson, I. 2001. "The Evolution of Modern Cadastres," accessed November 28, 2014, http://www.fig.net/pub/proceedings/korea/full-papers/session6/williamson.htm.

World Resource Institute. 2009. EarthTrends Environmental Information 2007, accessed August 17, 2009, www.earthtrends.wri.org/.

14

Displacement and Land Administration

Potel Jossam, Paul van der Molen, Luc Boerboom, Dimo Todorovski, and Walter T. de Vries

CONTENTS

Introduction

Serious conflicts tend to lead to both significant displacements of people and competing claims over land. Forced displacement, due to the involvement of arms, disrupts the relation that people have with their land, leaving them with no other choice than to leave their land behind for their own safety. However, the temporary disruption has long-lasting or even permanent effects in land tenure or even in the formal land administration as a whole. The vacated land is occupied by secondary and successive occupants, sometimes with the consent and under the direction of authorities. After the conflict, when original tenants return a conflict of interests emerges because of overlapping interests and conflicting claims, which may each be regarded as legitimate under successive administrations. On top of that, returning refugees—people who flee their homes for their safety and cross the border of their country—often find their original properties destroyed, leaving them with little proof or evidence to justify their claims.

The details of such scenarios have been described in a number of recent cases of conflicts. For example, in Mozambique, where the civil war started in 1977 and the conflict ended by 1992, secondary occupation was common, mainly by government employees, soldiers, and military officials (Todorovski et al. 2013). The conflict had displaced over five million refugees, whose original land claims could not be easily accommodated afterward.

Similarly, the various waves of ethnic conflicts in Rwanda in the period from 1959 to 1994 led to 2.5 million refugees and large numbers of internally displaced persons (IDPs) by 1994; at the same time, there has been massive numbers of returnees among those who had been refugees since 1959 (Potel 2014). Unfortunately, the list of conflicts and displacements is long. Kosovo, Cambodia, Sierra Leone, Liberia, Ivory Coast, Bosnia-Herzegovina, Sudan, Timor-Leste, Afghanistan, and Rwanda are the most reported ones (Zevenbergen and Burns 2010). All cases have, however, shown that acting on displacements and solving conflicting land claims from IDPs, refugees, and returnees is of crucial importance in supporting peaceful transactions away from the conflict. Land administration can thus be one of the supporting tools in post-conflict governance.

However, Todorovski (2011) argues that land administration in post-conflict periods is difficult to reestablish. It requires approaches that are able to deal appropriately with the circumstances of the local context and history. A crucial issue is, for example, the fact that often many of the original land records were destroyed, as was reported in the case of Timor-Leste (Todorovski 2011). In addition, former land officials were killed or displaced—the case in Rwanda—resulting in an abrupt and irreversible end of capacity in handling land issues. How can the land claims of displaced people and secondary occupants be reconciled responsibly, and in which periods of post-conflict can this be done and how? This is the main question addressed in this chapter.

This chapter will first address the theoretical views on how to regard the two main concepts displacement and administration of land and claims. This includes a short description of contemporary international views and policies on post-conflict administration. This is followed by a short introduction into the case of Rwanda, which is further explored through empirical analysis. After a description of the methodology to execute this analysis, a synthesis of the results and a discussion on the implication of these on each of the elements in the main research question follow.

Theoretical Perspectives

Two concepts need further theoretical elaboration: displacement and administration of land and claims in the post-conflict period. The term displacement refers first of all to the social context of forced migration by persecution or violence. In some cases however, it can also refer to development-induced displacement, that is, people find themselves forced to move because of lack of economic development (Alexander 2009). Yet, for the sake of this chapter we only refer to the first type of displacement. Displacement caused by especially armed conflicts always results in significant changes of land tenure, which remain largely undocumented due to the uncertainty in governance and administration that comes with conflicts in general (FAO 2005). Any occupation of the vacant land left by displaced persons is labeled secondary occupation, assuming that the original occupation was legitimate.

Displacement is an important concept in the aims of the so-called Pinheiro principles (COHRE 2003). These are nonbinding principles regarding housing and property restitution for refugees and IDPs to return to their homes and to recover their property. The principles were adopted in 2005 by the United Nations Commission of Human Rights. They provide a demanding set of rights that include granting the right of restitution for the lost homes and land to victims of dispossession with limited exceptions. Hence, on the issue of restoration after displacement the principles are unambiguous. They aim that all previous possessions and rights, thus including property and land rights, need to be restored in the manner in which they existed prior to the displacement. Though these principles are generally agreed on by most parties after conflicts, their implementation faces substantial challenges. Securing a durable solution of restitution of properties to all "rightful" owners is complicated as evidence for who is the rightful owner in which place is often lacking.

The second concept, administration of land and land claims or, better, reconciling and administrating conflicting land claims, is crucially related to displacement. Under ordinary circumstances, land administration refers to the processes run by government using public- or private-sector agencies to secure land tenure (usually by registering transactions and/or providing titles), determining and/or evaluating land value, land use, and land development. The underpinning land information system supports the land administration system, which provides an infrastructure for the implementation of land policies and land management strategies (Williamson et al. 2010). However, conflicts derive challenging circumstances, which first of all prevent the system from conducting administration activities in an ordinary way and secondly require the system to update itself in certain periods after the conflict.

The first difference with ordinary administration activities is the difference in both implicit and explicit legal consequences of granting land rights during periods of conflict and periods after conflicts. The ground on which a right is granted or transferred can be that a person loses any right of ownership after the expiration of a prescribed time of absence. In the case of a long-lasting conflict, the implicit legal consequence may be that people lose their right after 30 years or more of absence. This occurred in Rwanda, for example, where due to the conflict of 1959 to 1994 IDPs and returnees who had been absent for more than 30 years lost their legitimate right. As a result of this loss, the land was reallocated to the state. Still, the debate in this case was when one should start counting for calculating the 30 years. At the beginning of the conflict? At the end? Sometime in the middle? Or, when the country was stable and the displaced persons did not feel any threat to their life anymore?

The more explicit legal difference with administrating land in ordinary times is the fact of not having documented information at hand and/or not having formally licensed or legally recognized land administration staff members in times of conflict or immediately after a conflict. A crucial question in this is what can be considered documented evidence: evidence made by foreigners (not being part of the administrative system) or evidence based on witnesses (which cannot be verified). Often, as a result, administration can simply not take place in the ordinary fashion and legal shortcuts, in the form of discretionary executive decisions, need to be made. How this is done is thus a subject of empirical investigation of the present research.

In sum, the extent and type of displacement can be characterized by both the volume and proportion of primary and secondary occupations and the volume and size of restoration and restitution. Along with this, the administration of land and claims in post-conflict situations can be characterized by two types of evaluations: which legal principles are used to grants land rights and how executive decisions are made when documented evidence is absent.

A variety of literature sources report that both the displacement and the type of decisions on land claims differ in time passed during and after the conflict. Immediately after the end of the conflict, many institutions are not yet functioning normally and therefore normal administration and enforcement by the state institutions and avoiding illegal occupation are not possible (iDMC 2013). Thus, the post-conflict situation needs to be further detailed based on what is functioning and what is not. Three types of post-conflict periods are earmarked in the literature:

1. The *emergency* period. This is the period in the immediate aftermath of the conflict before full-scale mobilization of aid resources has started (UN/HRIDP 2008). During this period, emergency activities focus on establishing basic governance and providing humanitarian services (UNHCR 2010), as there is often little or no operational governance and rule of law and there is extensive destruction of infrastructure.
2. The *early recovery* period. This period is that transitional phase of the post-conflict country in which legitimate local capacities emerge and should be supported with particular attention needed for restarting the economy. This includes physical reconstruction ensuring functional structures for governance and judicial process and laying the foundations for provisions of basic social welfare such as education and health care, and hence social stability (Nkurunziza 2008). This period involves the development of a legal framework, national policy developments, state formation, and developing strategies for their implementation.
3. The *reconstruction* period. In this period, the rebuilding of a society and physical infrastructures proceeds (van der Molen 2004).

> Government in a post-conflict situation is likely to need revenue and land tax can be an important source (UN-HABITAT 2007), and for this to succeed land in a country needs to be well administered. In the reconstruction period, much attention focuses on the implementation of policies. In land policy, implementation of developed policy on access of land should be supported by land legislation, adjudication procedures for land claims and disputes, existing land administration systems, housing strategies, eviction procedures, administration of state-owned land, administration of private abandoned land, and transparency (FAO 2005).

Hence, when evaluating the relation between displacement and administration of land and claims, the characteristics of both need to be taken into account within the aforementioned post-conflict periods.

The present research evaluates the relation for the specific case of Rwanda. Rwanda faced two violent conflicts rooted in ethnic differences between Tutsis and Hutus. It caused enormous displacement over a large period of time and many land claims. In 1959 a large number of Tutsis fled, and in 1994 the Hutus fled (in fear of retribution). During the entire period from 1959 to far into the twenty-first century, displacements continued, resulting in secondary occupation. In conflict, many IDPs and refugees are forced to flee and leave behind their land without legal documents that can later be used to justify their ownership (Augustinus et al. 2004); this was the case in Rwanda. According to a UNHCR (2010) report, Rwanda had over half a million refugees in neighboring countries by 1964.

In 1959, the Tutsis were forced out of the country, yet they could not justify their land rights as the country had a weak land administration system at the time. As a result, in 1962 the land that had belonged to those Tutsis was distributed to Hutu peasants (Potel 2014). With the Rwandese Patriotic Front (RPF) taking over power after the worst genocide and violence in 1994, there were over 2.5 million refugees and IDPs. After this, a massive return of the first wave of returning refugees started, resulting in massive conflicting land claims (Potel 2014). By 1994, Tutsi refugees had returned and Hutu had fled. Tutsi returnees found the land to be reoccupied since Hutu had left (Jones 2003). However, in the years 1996–1997, when the country was stable, more than one and a half million Hutus returned to Rwanda, only to find that the land they had been occupying for more than three decades had been returned to the returning Tutsis (Potel 2014).

By the beginning of 1997, the country was facing the challenge of resettling the returning refugees (Murekezi 2012). Thus, dealing with land claims and disputes was placed high on the political agenda. Despite the lack of a legal framework, a number of measures were taken to achieve sustainable peace and stability and meet the land-related claims and disputes. These measures were land sharing, village development processes (*Imidugudu* settlement), and a new land policy leading to countrywide land reform especially the part of Akagera National park. In embarking on major land reform, the government wanted to develop proper land governance that would enable the population to enjoy a more secure form of tenure and bring about proper land utilization, ensure efficient land management, generate taxation income, and regulate the land market (COHRE 2003). However, during the implementation more challenges related to claims and disputes arose during initial land registration, and this ensued from the way how government handled post-conflict land issues.

Methodology

To understand both displacement and land claims in three different periods after a conflict, the empirical part of this research aimed at collecting the proxies of both in these three periods. The data collection relied on a combination of both primary and secondary information sources collected during a fieldwork. Primary data were collected through interviews, focus group discussions, and field observations. Secondary data were collected from governmental organizations at sector level, mediation committees, the district office, and the Office of Ombudsman, the Rwanda Natural Resource Authority, and the National Secretariat for Mediation committees.

All of these were collected in the eastern province of Rwanda, in the districts of Kayonza and Ngoma. These areas were particularly affected by massive numbers of returnees from Uganda, Tanzania, and Burundi. Because of this massive land need, there was a high degree of land sharing.

After the consultation of district officials on access to the sectors and population, two sectors were chosen, a sector per district where interviews were carried for data collection. Thus, in Kayonza Mukarange sector was chosen, whereas in Ngoma district Remera sector was chosen. Further, a stratified random sampling method was also used to choose respondents for four target groups: the 1959 refugees, 1994 refugees, IDPs, and survivors of the Tutsi genocide.

For the primary data collected, interviews, focus group discussions, and field observations were employed. Interviews were conducted with 31 respondents from each of the 4 targeted groups. The focus group discussion involved 12 former leaders, representing each of the 4 targeted groups as well. The focus group discussion was executed in 2 different sessions with 6 participants each, whereby the discussion focused on how they handled land claims and disputes of returnees in the post-conflict period. As the research was intended to know how land claims and disputes were handled in post-conflict Rwanda during emergency and early recovery periods, additional related questions were asked. Questions like how repossession of land was possible and how land sharing policy and Imidugudu policy were implemented were additionally asked to moderate the interaction of the fieldwork discussion. Finally, field observations were collected throughout the fieldwork period.

The secondary data collection relied on assembling reports and government documentary evidence on how land claims were addressed by governmental institutions in each of the periods. These documents were collected from various sectors, mediation committees, the district office, the Office of Ombudsman, the Rwanda Natural Resource Authority, and the National Secretariat for Mediation committees.

Results

The data are grouped according to (1) post-conflict period types and with respect to the issues of (2) displacement [characterized by (1)], and they are presented in the following structure: volume and proportion of primary and secondary occupation, volume and size of restoration and restitution of land claims [characterized by (1)], legal principles to grant land rights, and the manner in which executive decisions are made.

Emergency period: regarding the volume and proportion of primary and secondary occupation in the immediate aftermath of the 1994 war, documents and individual accounts suggest that the vast majority of the 1959 refugees returned to the land they claimed to have previously owned and thus reoccupied the land that had been left by the fleeing Hutus.

Although the exact volume and size of restoration and restitution remains unknown in this period, in 1994 the authorities decided to locate the returning Tutsis to houses and fields that had been abandoned by the fleeing Hutus as an immediate solution to the challenge of housing and land for agriculture because administrative authorities were not in a place to provide them in such time.

As far as the legal principles to grant land rights are concerned, in 1962 the government of that era distributed the land of the displaced Tutsis to the local Hutus in a political move commonly known as "land for democracy" (a common word used to refer to a move used by the government authority in 1962 during the distribution of Tutsi land to the Hutus, claiming to be sharing the fruits of the revolution by the Hutus against the Tutsis). Immediately after the war, there was no specialized government body in charge of land; all local authorities were regulating all issues, including land issues, with no formal guidelines to follow while dealing with land claims and disputes in case they arose. The legal principles were highly influenced by pragmatism to overcome immediate social and political challenges and focus on food, shelter, and security.

It is clear from most respondents that the manner in which executive decisions were made was largely ad hoc. The majority of the interview respondents mentioned that in 1994 there was no concrete structure or plan regarding land management as such. The primary attention was to secure food and safety of the returning refugees. So, there were initially no complaints about land allocation in 1994. One local leader specifically remarked, "By that time, we (local authority) were not concerned with land issues, what mattered to us as authority was food and shelter for returning refugees, and because of that, some individuals used that governing crisis and went on to grab land that was left vacant by 1994 refugees. By this time, the challenge the government had was not land issue, what was so challenging was to let these returnees have food and shelter" (Potel 2014).

Early recovery period: as the conditions of life within the country started to improve, security was calm, displaced people had returned, no more emergency needs arose, and the actual volume and proportion of primary and secondary occupation decreased considerably, but the reported cases to handle seemed to increase dramatically. The explanation can be found in the fact that the main emphasis now shifted to granting restoration and restitution. Since the Arusha peace agreement prohibited the 1959 refugees from repossessing the land they had previously owned, except if their houses still existed, the respondents mentioned during focus group discussions that there was a challenge of providing land access for these returning refugees and IDPs who could not return to their place of origin. One of the challenging issues was the restitution of 1959 returnees who stuck to their land on the basis of the Arusha peace agreements. The solution to this challenge for the government was to use one part of the Akagera National Park as restitution land. This was a way of getting land for the 1959 returnees who were not going to reoccupy their former land.

Regarding the legal principles to grant land rights in the early recovery period, a settlement policy commonly known as Imidugudu settlement (Imidugudu or Umudugudu for singular is a grouped or collective settlement) was most prominent. The administrative regulations regarding land sharing was later too be referred to as the land sharing policy (the policy required a returnee, either of an IDP, or a refugee of 1959, had to share the land with a returnee of 1994, or a secondary occupant on the land he initially owned before 1959). The development of the Imidugudu policy was an effort by the government to tackle provisioning of emergency shelter and housing to homeless people, development of a housing strategy for returning refugees and IDPs whose homes and houses had been destroyed, provisioning of access to land by returning IDPs and refugees, prevention of illegal occupation of land by non-beneficial or illegal occupants, restitution of land to its "lawful" owners by the establishment of mechanisms to resolve land claims and disputes, and finally establishment of fair procedures for eviction of unauthorized occupants (FAO 2005).

As far as the manner in which executive decisions were made is concerned, it is obvious from the personal accounts that the attention of the authorities and nongovernmental organizations shifted from carrying out humanitarian emergency activities to the development of policies aimed at transforming the country toward recovery. The land sharing policy required two belligerent persons (a Tutsi who owned a land before 1959 and a Hutu who occupied the land since 1959 to 1994) to share that land. The Hutu here was referred to as the principal owner and, thus, given the right to choose which side of the land to take, and the Tutsi was the beneficiary. The Imidugudu policy required all people to build and settle in a grouped settlement (nuclear settlement). The policy provided that the Imidugudu land belonged to state and, therefore, plots were given to people by the state to build on and, thus, a private person would build a private house on state land. Still, as Potel (2014) noted, this land sharing policy was not uniformly implemented all over the country and there were no formal regulations establishing it.

Reconstruction period: regarding the volume and proportion of primary and secondary occupation, it can be said that due to the fact that implementing land reform based on formal land policy and land law was top priority there was no increase in secondary occupation. Instead, authorities faced the challenge of handling the continuously increasing land claims and disputes. These claims were arising from how land issues were handled during the emergency and the early recovery period onward. Many land issues were more about family-related issues (complaints between family members) but with the roots of what happened during land sharing (Potel 2014). A specific case reported by one of the interviewees describes this best:

> Mr. Black* left the country in 1959 after the massacre of his father and grandfather. They had 30 hectors of land with a forest and a house in it, and the land was redistributed to other local people by the then government in 1962. Black returned back in 1994, and found that those who have been occupying the land had also been displaced. Mr. Black re-occupied his land, during 1997 land sharing; he opposed the sharing of that land with anybody claiming it to be a family land to all descendants of his late grandfather Late Brown. As the country regained stability many of the people who had acquired and occupied that land returned and started to lodge claims to the authority over the right to the same land. In an attempt to solve the issue by 2009, there were 37 claimants, all claiming rights to that land.

As far as the volume and size of restoration and restitution is concerned according to the general report of the Land Tenure Regularization Program (LTRP), a nationwide program aimed at registering land all over Rwanda, 1,494,943 parcels were not registered (14% of demarcated land) and under dispute. Information for these parcels was incomplete due to the fact that the

* Not the real names; names anonymized.

true owners either had been killed and had no next kin to take over the lands or were displaced and had never returned. Additionally, 0.13% of demarcated parcels with complete information is under dispute and has thus remained unregistered as per LTRP regulations. Land that was under claims or disputes during the LTRP process was skipped in the regular registration process and registered instead in a claims registry. Claimants were thus directed or referred to competent authorities or court. If a decision was taken, then such a parcel would be registered again with the true owner. As stated in the final report of systematic land registration (SLR), some parcels were recorded as having incomplete information; this was due to the fact that the true owner either was dead (most of the people died during the genocide conflict) and had no next kin to reclaim the land or was displaced and has never returned back with no next kin to take care of the parcel and other parcels are registered under disputes. All these have hampered land administration as the said parcels have remained with no land certificates issued and, thus, the government's goal of registering and issuing land certificates all over the country for proper land administration is hampered by displacement.

On the issue of legal principles to grant land rights, it was noted that although Imidugudu eased the perception of people toward Imidugudu and encouraged them to join the Imidugudu, it created a challenge later during the LTRP. The Imidugudu policy provided that the umidugudu land belonged to state and therefore plots were given to people by the state to build on and thus, a private person would build a private house on state land. Although this eased the perception of people towards imidugudu and encouraged them to join the imidugudu, it created a challenge later during Land tenure Regularization System Programme (LTRSP). The reconstruction period in Rwanda brought some remarkable changes in matters concerning land as there was a new land policy, the constitution, and the organic land law. Bruce (2007) outlines different issues tackled by the new land policy and the land law, among which are the following:

> Customary holdings are recognized, but they are to be converted to leaseholds held from the state and abolish the customary tenure and continue to cancel all occurrences of customary feudalism within the country.
>
> Registration of land became mandatory.
>
> The state is responsible for giving land to persons who were denied their property rights.
>
> Land sharing that has occurred since 1994–2005 is explicitly validated, and those who received land are recognized as having the same rights as other customary holders.

The law recognizes that the 30-year occupation of land by a private user or the state gives rise to rights for leasehold but excludes occupations initiated by violence or fraud and makes it clear that what is acquired is not ownership but leasehold right on state.

On the manner in which executive decisions were made, it is obvious that handling the claims remained complex even during this period. Interviewees indicated that many of the complaints were overlapping or interrelated, referring to either the Tutsi–Hutu conflicts of 1959 and 1994 or the land sharing of 1996–2003. Others were related to genocide survivors' land, especially orphans' land occupied by others or by the state; others were related to family members due to how land sharing was handled; and others were related to land taken by Imidugudu settlements.

Discussion

The conflicts in Rwanda that resulted in the 1959 Tutsi displacement and the 1994 displacement due to the genocide against Tutsis affected the social and economic life of the people in particular and the country in general. This created a challenge of handling successive and conflicting legitimate rights, which ultimately can become a backbone of future land administration. How can one understand these experiences in the broader light of displacement and administrating land claims in post-conflict periods? And, which characteristics are fundamentally different in the emergency, early recovery, and reconstruction periods, respectively?

First of all, the volume and proportion of primary and secondary occupation clearly differ per the post-conflict period. In the emergency period, many of the figures tend to be undocumented, although the actual displacement may be high. Yet, especially during the early recovery period the statistics tends to rise enormously, as people become more aware of the ability to claim land on formal grounds.

Second, to deal with the volume and size of restoration and restitution the acting government of Rwanda in the different periods adopted different strategies. Especially due to the multiple conflicts and displacements over time, the ruling administration tended to recognize the secondary occupation only. Implicitly, it would seem therefore that this restoration relied on a system of land administration that guaranteed only secondary occupation claims in favor of primary occupation claims. Although there have been some critiques of the approach of Rwanda to deal with restoration in view of land claims (Jones 2003), international instruments such as the Pinheiro principles were not yet in place at the time. The Pinheiro principles were adopted in 2005 when many of the Rwandan land issues had already been dealt with. In addition, it is unclear in retrospect whether the application of these principles would have solved the Rwandan challenges; instead, they would have aggravated the complexity of the challenges because the Pinheiro principles advocate the restoration of things as they were before displacement. With reference to the example of Mr. Black's case, allowing 1959 returnees to reoccupy his former land as the Pinheiro principles suggest would have left 37 families landless. Given the fact that there were similar cases, implementation of these principles would have caused more challenges to the country than solutions.

Regarding the legal principles to grant land rights, various frameworks and references have been applied in each of the post-conflict periods. The example of Mr. Black showed that initially the Pinheiro principles were implicitly applied. The argument was related to the issue of land scarcity in the country, whereby many of these claimants were landless yet entirely dependent on agriculture (subsistence farming). Based on this argumentation, any claimant would deserve the right to be reinstated on land and then the government had to get land for the Hutu returnees and settle them there. However, in Rwanda's context it is hard to say who needed restitution, or who needed compensation, taking into consideration that Mr. Black's family was forced to leave the land through forced displacement and, by effect, in legal matters no prescription that would have applied, he had found his father's house and grandfather's forests, so he had all rights to repossess the land. On the other side, however, the ruling authority in 1962 had legitimized the occupation of abandoned property, meaning these people also had a legal claim and thus deserved legal protection.

Another legal framework was the Arusha agreement. This agreement still applied in both emergency and early recovery periods. The agreement provided that "in order to promote social harmony and national reconciliation, refugees who left the country more than 10 years ago, should not reclaim their properties, which might have been occupied by other people. The government shall compensate them by putting land at their disposal and shall help them to settle." The agreement went against international norms, or the Pinheiro principles (i.e. that all claims to previous rights should be respected and available to restitution), however, pragmatism saw the 10-year time frame adopted.

Finally, the Imidugudu policy and the LTRP supported the reconstruction as legal and regulatory frameworks. Especially in the reconstruction period, which can be characterized by essential political stability, more such elaborate programs were possible. The type of legal principles draws in this period on a larger degree of stakeholder consensus and a smaller degree of administrative routine.

Regarding the manner in which executive decisions are made, it is fair to conclude that the type of executive decisions draws gradually on a lower degree of discretionary and ad hoc decisions and a higher degree of decision types that are based on administrative routines. In an emergency post-conflict period, there is no other option than to make decisions in an ad hoc way, based on what is considered appropriate. Given the number of people who had returned, and the post-conflict situation the government was handling, they would neither get the manpower to do the work nor would the approach solve the land problem and challenges related to land that people had. Some of them did not even remember their place of origin, especially the 1959 refugees; there were no land records that would prove who were the true owners since there was no formal record and restitution would lead many people to become landless. Pinheiro principles advocate individualized property restitution, which is return to each of the particular lost property. In Rwanda's case, this would pose a number of challenges for administration and implementation, as individualized restitution requirements would create the problem of fairness due to a number of claims because restitution would only be accomplished through separate individualized proceedings.

In the early recovery period, the government focused on land issues and to deal with the challenges in the land sector that involved restitution of land occupied by others, the government adopted two policies: the land sharing and Imidugudu policies. During the land sharing process, the Hutu who had acquired land during the earlier period, were considered principal owners. Tutsi returnees were discouraged from repossessing their former land by the Arusha peace agreement whilst the discouragement of claims can be criticize as it contradicts international guidelines regarding housing, land, and property rights, arguably it greatly reduced the number, degree, and overlapping of claims and disputes. However much discouraging the Tutsi returnees from repossessing their land can be criticized for contradicting the international guidelines regarding housing, land and property rights and Pinheiro principles, of which Rwanda is a party, it reduced the number and degree at which such claims and disputes were increasing, since few 1959 returnees remained with such claims as much as whilst the two policies did succeed in stabilizing the country and reducing suspicion and mistrust among Rwandans, transferring the policies to another country would require a thorough study of the history and sociocultural context.

It was only in the reconstruction period that the government of Rwanda could identify land as a very important element for the future development of post-conflict Rwanda. The land issue was very high on the political agenda with a nationwide land reform, and since then a lot of developments in legal, institutional, and organizational aspects have been recorded. There is a national SLR program aimed at registering all the land all over the country; although the program had been applauded for its success, it faced some challenges of recording a number of lands under disputes and others whose fate was not yet determined due to insufficient information. Consequently, efforts to deliver a functioning nationwide land administration system, as required by the national land policy, have been hampered. The Rwandan experience makes it very clear that post-conflict countries should focus early on formalizing land administration: it brings the key issue of land tenure security to the fore, as is usually necessary, whilst supporting revenue raising for post-conflict reconstruction.

Conclusion

As the research question was aimed at knowing how land claims and disputes were handled in post-conflict Rwanda and how this later affected land administration, questions were drawn on how land claims and disputes were addressed in emergency and early recovery periods of post-conflict Rwanda and who the actors were in dealing with land administration in these aforementioned post-conflict periods. In the emergency period of post-conflict Rwanda, there was a total loss of focus on land claims and disputes; due to political and social issues, the post-conflict government had to deal with issues such as security and the number of returnees (for their food and shelter). In the early recovery period, the government turned its focus on land claims and disputes. To deal with the challenges in the land sector that involved restitution of land occupied by others, the government adopted two policies, mainly land sharing and Imidugudu policies; for reaching a sustainable solution on the land issue, the government of Rwanda started a new nationwide SLR program aimed at registering land all over the country. Although the program had been applauded for its success, it faced challenges in recording lands under dispute, sometimes in a very information poor environment.

The case of Rwanda is partially exemplary for other post-conflict cases, where both displacement and administrating land claims occur. It seems likely that in post-conflict periods, where the volume of displacement and the proportion of primary and secondary occupation are severe, the process of establishing an administration responsibly regarding land claims, can only shift gradually from an ad hoc to a more sustainable codified approach. Establishing normative frameworks and associated implementation strategies from the onset is likely to result in extensive administration requirements, which cannot be responded to in the earlier post-conflict periods. Moreover, it remains very difficult to generalize land issues and administrative requirements, as each post-conflict society has its unique sociocultural setup.

References

Alexender, B. 2009. *Forced Migration and Global Politics*. West Sussex, UK: Wiley.

Augustinus, C., D. Lewis, J.A. Zevenbergen, P. van der Molen, and T. Naudin. 2004. *Land Administration: Handbook for Planning Immediate Measures from Emergency to Reconstruction: Peer Review Draft 1*. Nairobi, Kenya: United Nations Human Settlements Programme (UN-HABITAT).

Bruce, J. 2007. Drawing a line under the crisis: Reconciling returnee land access and security in post conflict Rwanda. London: Humanitarian Policy group, Overseas Development Institute.

COHRE. 2003. *The Pinheiro Principles: United Nations Principles on Housing and Property Restitution for Refugees and Displaced Persons*. Geneva, Switzerland: Transnational Publishers.

FAO. 2005. *Land Tenure Studies: Access to Rural Land and Land Administration after Violent Conflicts*. Rome, Italy: FAO.

iDMC. 2013. *Global Overview 2012: People Internally Displaced by Conflict and Violence*. Geneva, Switzerland: Norwegian Refugee Council/Internal Displacement Monitoring Centre (NRC/IDMC).

Jones, L. 2003. Giving and taking away: The difference between theory and practice regarding property restitution in Rwanda. In Scott Leckie (ed.) *Returning Home: Housing and Property Restitution Rights for Refugees and Displaced Persons*. New York: Transnational Publishers.

Murekezi, A. 2012. *Rebuilding after Conflict and Strengthening Fragile States: A View from Rwanda* (e-book). Vol. 22, ACBF Working Papers. Harare, Zimbabwe: African Capacity Building Foundation (ACBF).

Nkurunziza, J.D. 2008. "Civil War and Post-Conflict Reconstruction in Africa." United Nations Conference on Trade and Development (UNCTAD), Geneva, Switzerland.

Potel, J. 2014. "Displacement and Land Administration in Post-Conflict Areas—Case of Rwanda." Master of Science in Geo-information and Earth Observation, Land Administration, University of Twente, Enschede, The Netherlands.

Todorovski, D. 2011. "Characteristics of Post-Conflict Land Administration with Focus on the Status of Land Record in Such Environment." FIG Working Week 2011: Bridging the Gap between Cultures: Technical Programme and Proceedings, Marrakech, Morocco, May 18–22, 2011. p. 14.

Todorovski, D., J.A. Zevenbergen, P. van der Molen, and E.M.C. Groenendijk. 2013. "Post-Conflict Land Administration and Its Current Status as Facilitator of the Post-Conflict State Building: Case Mozambique." Proceedings of FIG Working Week 2013, Environment for Sustainability, Abuja, Nigeria, May 6–10, 2013. Copenhagen, Denmark: FIG, 2013. ISBN: 978-87-92853-05-9. p. 11.

UN-HABITAT. 2007. *A Post-Conflict Land Administration and Peacebuilding Handbook*. Nairobi, Kenya: UN-HABITAT.

UNHCR. 2010. *Handbook for the Protection of Internally Displaced Persons, Global Protection Cluster*, Geneva, Switzerland: UNCHR.

UN/HRIDP. 2008. *Protecting Internally Displaced Persons: A Manual for Law and Policymakers*. Bern, Switzerland: University of Bern, Brookings Institution.

van der Molen, P. 2004. "Land administration in post-conflict areas: Editorial." *Journal Abroad: Periodical Newsletter of Kadaster International* 8 (2):3.

Williamson, I., S. Enemark, J. Wallace, and A. Rajabifard. 2010. *Land Administration for Sustainable Development*. Redlands, CA: ESRI Press.

Zevenbergen, J.A., and T. Burns. 2010. "Land Administration in Post-Conflict Areas: A Key Land and Conflict Issue." XXIV FIG International Congress 2010: Facing the Challenges: Building the Capacity, Sydney, Australia: Technical Programme and Proceedings, April 11–16, 2010. p. 22.

15

Social Tenure Domain Model: An Emerging Land Governance Tool

Danilo Antonio, Jaap Zevenbergen, and Clarissa Augustinus

CONTENTS

Introduction

A key component of land administration is a reliable and affordable land information system. Nevertheless, estimates on land information coverage, cadastres, or land registries suggest that in developing countries 70% of the area is not covered by any formal land registration system (Lemmen et al. 2009; Lemmen 2010; United Nations Human Settlements Programme/Global Land Tool Network 2012; Zevenbergen et al. 2013). Enemark et al. (2014) even argued that 75% of the world's population do not have access to formal land rights systems. The majority of these people are the poor, women, and vulnerable groups who have very limited access to land (Zevenbergen et al. 2013). This is especially prevalent in large informal settlements and slums that result from rapid urbanization processes without the presence of a formal land sector ready to deal with that. In this regard, the United Nations Human Settlements Programme (2008a) reports that the global slum population is expected to increase to 1.4 billion in the coming years.

Land information is important for the promotion of good land governance that contributes to poverty reduction and sustainable development objectives. Information empowers people and thus strengthens democracy. It is strategic to bridge the information gaps that exist between the various types of informal land tenures and methods of officially recording them. Although some debate still continues in specific academic and professional discourses, there is increased acceptance of the need for diversity in people to land relationships. This range of land tenure realities became known as "the continuum of land rights" within the United Nations Human Settlements Programme (UN-Habitat) and among the other partners of the Global Land Tool Network (GLTN) and has received increasing buy-in in recent years. Accordingly, tenure can take a variety of forms and one can view rights to land as lying on a continuum between formal and informal rights. In between these two extremes, there are a wide range of rights; however, the rights do not lie on a single line, and they may overlap with one another (United Nations Human Settlements Programme 2008b; United Nations Human Settlements Programme/Global Land Tool Network 2012).

Acceptance of this reality led to the need to design and implement innovative land information tools to record these rights. The Social Tenure Domain Model (STDM) is an important tool designed for this, but it has emerged as a much broader land tool with the potential to contribute to improved land governance in several ways. After STDM was initially developed and refined, its first pilot implementation took place in 2011–2012 in Mbale municipality in Uganda.

In this chapter, we analyze STDM and the lessons from this pilot project against its original land recording design and its broader land governance roles to understand why it is now considered as a key land tool within the broader GLTN discourse. We start with a short theoretical overview of some key land governance challenges, as well as some conventional and innovative domain models (Land Administration Domain Model [LADM] and STDM). With these key elements in mind, we present the components of the pilot design and its results, as well as the lessons learned. Finally, we come to the discussion of its roles as a key land tool within the broader global discussion of pro-poor land tools and the conclusions.

Theoretical Perspective

"Land governance" is defined by UN-Habitat and the Food and Agriculture Organization (FAO) as "concerning the rules, processes and structures through which decisions are made about access to land and its use, the manner in which the decisions are implemented and enforced, the way that competing interests in land are managed" (Palmer et al. 2009). They emphasize that land governance embodies policy, legal, and institutional frameworks surrounding statutory, customary, and informal land practices and transactions. Further, they rightfully focus on the analysis and emphasis of "power" and "politics" surrounding land management and administration.

Providing tenure security to all members of the society is a key indicator of good land governance. A wealth of literature emphasizes the need for security of tenure and elaborates on its benefits, which include contributions toward poverty reduction, sustainable development, management of land disputes and conflicts, improvement of land use planning, management of natural resources, and protection of the environment (Antonio 2006; Augustinus 2009; Burns 2007; Deininger 2003; Deininger and Enemark 2010; Food and Agriculture Organization of the United Nations 2007; Magel 2006; Palmer et al. 2009; United Nations Human Settlements Programme 2008b; United Nations Human Settlements Programme/Global Land Tool Network 2012; Williamson et al. 2010). The literature also stresses that security of tenure is the gateway to other significant processes and opportunities for poverty reduction and sustainable development.

The following challenges need to be overcome to deliver secure tenure to all and thus improve land governance (Augustinus 2009):

1. Underrepresentation of women, who make up 50% of the world's population but only own 2% of the world's land.
2. Lack of political will by governments to go to scale and address the needs of the whole population.
3. Proliferation of graft and corruption in the land sector. Transparency International (2009) reports that the land services sector is the third most corrupt sector after the police and the courts and that petty corruption in this sector costs USD700 million in India alone (Transparency International-India 2005).
4. Incompleteness and unreliability of land records and land information systems that are not interlinked in most countries.
5. Lack of sufficient pro-poor and practical land tools and approaches to assist the implementation of land policies.

Innovative land information tools like STDM at first sight address the fifth challenge; but when implemented in a participatory and transparent way, within a good governance framework, they help to deal with more challenges. With all conditions in place, one could theoretically assume that they would contribute to improvements on all the five challenges. A key provision is that they should support and allow the implementation of a continuum of land rights approach.

"Data domain models" seek to abstract reality into conceptual entities, characteristics, and relationships. In this regard, although STDM currently stands for a way of thinking, a software package, and a data collection approach (United Nations Human Settlements Programme/Global Land Tool Network 2012), the last two words of its name ("domain model") show its roots as part of a data modeling exercise that started as the Core Cadastral Domain Model (van Oosterom et al. 2006). With the support of the International Federation of Surveyors (FIG) and the International Organization for Standardization (ISO), this led to the international standard 19152 called the Land Administration Domain Model (Lemmen 2012). LADM is primarily intended as a reference standard for conventional land registration and cadastres. Its roots go back to the (Western-style) core entities: owner, right (title), and parcel (e.g., Henssen 1995). In LADM, this is already generalized to the basic classes party; rights, restrictions, and responsibilities (RRR); and spatial unit. However, to handle the broad range of land tenure options that can be found among many, especially poor, people, more flexible notions are necessary, such as the continuum. It is therefore necessary to not only interpret party and spatial unit more widely (including, e.g., groups of groups and "dots-for-plots") but also rename the relation between these two as social tenure. This is also where the name "Social Tenure Domain Model" finds it roots. Social tenure also implies that it is about not only legal but also nonlegal tenure types, which are the social relationships between people and land.

In summary, both LADM and STDM represent the relationships that people have with a given piece of land or spatial units.

Building from Lemmen (2012), the core of LADM is a conceptual model that provides an abstract representation of the following:

1. Party: This can be parties (people or organizations) or group parties (groups of people or organizations). A party constitutes an identifiable single entity that is legally recognized.
2. Ownership right: This is a formal or informal entitlement to own a given piece of land. This can also represent a responsibility (e.g., to maintain a historical site or a waterway) or a restriction over a given piece of land (or water).
3. Spatial unit: This represents a single area (or multiple areas) of land (and/or water) to which a party has a right, restriction, or responsibility.
4. Spatial sources and representations: This refers to the documents related to a survey in which the data are acquired digitally in a field office, using forms of field sketches, ortho-images, or existing topographic maps.

By abstracting the representation of the basic components of land administration, LADM provides common terminology for land administration. The application of relevant attributes, rules, and associations to these core components enables LADM to provide a basis for national and regional profiles that can be subsequently standardized and easily integrated.

In this manner, STDM can be seen as an implementation of LADM, specifically for developing countries that have very little or no cadastral coverage in urban or rural areas. STDM, in its current form, tries to represent the realities on the ground, particularly modeling the complex tenure relationships that poor people have, independently of accuracy or legal issues.

STDM represents the basic concepts in three ways:

1. Party: This can be a person, company, municipality, cooperative, married couple, group, or group of groups.
2. Social tenure relationship: This is the relationship between parties and spatial units and is in the form of informal rights, tenure rights, long leases, rents, ownership rights, Islamic tenure rights, state property, conflict areas, disagreements, overlaps, use rights, and so on.
3. Spatial units: Discrete areas of land, natural resources, properties, structures, or objects other than accurate and well-established units (relative to those defined by LADM).

STDM also envisions capturing supporting documents, including photographs, images, videos, sketches, maps, and other documents (formal or informal), to illustrate and describe the social tenure relationships and to identify the concerned parties. Figure 15.1 represents the STDM conceptual model.

The commonalities between LADM and STDM include the following:

1. Rights, restrictions, or responsibilities in LADM are collectively represented as a social tenure relationship in STDM.
2. The right, restriction, or responsibility (in LADM) and social tenure relationship (in STDM) of a party over a spatial unit can be supported by both source documents and supporting documents.

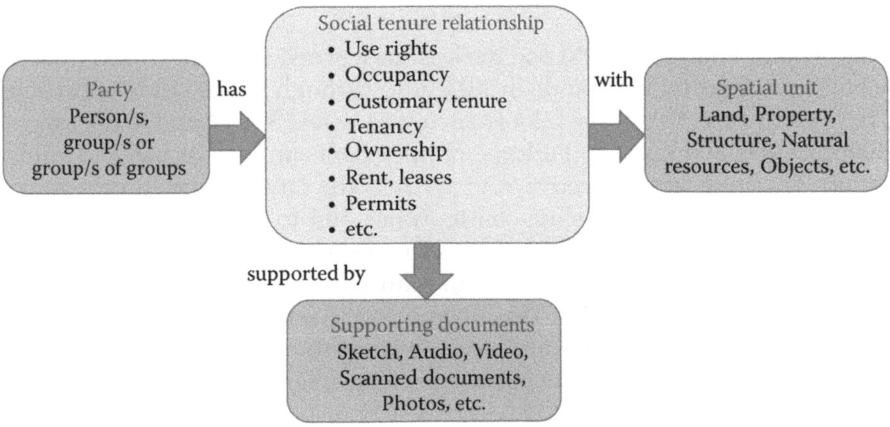

FIGURE 15.1
The Social Tenure Domain Model conceptual model. (From Antonio, D. et al., Addressing the Information Requirements of the Urban Poor—STDM Pilot in Uganda: A Brief, UN-Habitat/Global Land Tool Network, Nairobi, Kenya, 2014.)

However, in STDM this is expanded to informal documents or even taped verbal witnesses. It can include these because it is not limited or tied to representing a formal or legal tenure relationship.

3. Both LADM and STDM support state-based modeling in which the validity of objects requires at least two dates (start date and end date) to indicate the time interval when the object is actually recorded in the system. Consequently, it is possible to reconstruct the particular state of an object for any moment or period in time.

The main difference between LADM and STDM arises from their respective origins and context. Whereas LADM refers to the general context and seemingly highlights more its use in formal systems, such as cadastre and land registry systems, STDM extends this to a specific concept. STDM further emphasizes the relationships between people and land (extending its definition to natural resources, buildings, properties, and other objects), independently of the level of formalization, involved technical accuracy, and legal status of these relationships. Griffith-Charles (2010) states that STDM is a more flexible way of representing the actual tenure arrangements and further argues that LADM is too general to describe the nuances of many existing informal and non-standardized tenure relationships. The two models are complementary in most cases, particularly in situations where conventional cadastral systems cannot support tenure security, such as post-conflict/disaster situations, informal settlements, and customary tenure areas.

Methodology

This chapter analyses STDM and the lessons learned from the STDM pilot in Mbale municipality in Uganda in 2011–2012 through a case study approach. The specific objective of the pilot as such was to test STDM and to document the processes and capacity building requirements around its use and application. The long-term objective is to address the land information requirements of people living in slum communities and to build their capacity in the use and application of the land information systems based on free and open source software packages and in mainstreaming the thinking behind the continuum of land rights. This is expected to form the basis for dialogue between local communities and cities on improving tenure security and inclusive planning and enhancing access to basic services and infrastructure (Antonio et al. 2014).

The analysis that this chapter undertakes can be seen as an add-on to this objective at the next level of abstraction. In addition to the pilot documentation, the first author's involvement adds information to the case study. The Mbale pilot constitutes a case study of not only the use of STDM as a land (information) tool but also the way in which the partnership to implement the pilot built on a variety of contributions from different partners to make it work. Although the impact on the landholders of the two slums is the key output from the use of STDM, the impact for other partners at different levels is also taken into account.

The project was implemented alongside an existing larger project called Transforming the Settlements of the Urban Poor in Uganda (TSUPU). This project aimed to reenforce linkages between various urban sector programs and initiatives by introducing systemic changes in delivering urban services, improving urban management, and developing planning and policy frameworks. The government of Uganda, through the Ministry of Lands, Housing and Urban Development (MoLHUD), leads the project. The Municipality of Mbale was chosen as the pilot area. The project design was based on two pillars: strengthening partnerships at all levels and building on community strengths and processes. The Cities Alliance, Slum Dwellers International (SDI), UN-Habitat, GLTN Secretariat, MoLHUD, ACTogether, and Mbale municipality provided advisory, technical, and capacity development inputs. Most of the work, including mapping, enumeration, community mobilization, local consultations and sensitization and data entry, and analysis and validation, was done by the Mbale Slum Federation and community members.

An analysis of the pilot (case) is presented along the lines of partnership, pilot activities, and land information and its use, all within the objective given at the start of this section. The section ends with an overview of the impact and lessons learned.

Results

A "partnership" was provided as the underlying platform on which the piloting commenced. With the large number of partners involved in the pilot, it became clear that three dominant types of partners could be distinguished based on the roles and responsibilities that they contributed to the pilot: facilitators, supporters, and implementers. Antonio et al. (2014) describe the specific roles, and these are not summarized.

The *facilitators* included Cities Alliance and the FIG. In addition to being the funding agency for the project, Cities Alliance strategically placed the project to coincide with the implementation of the TSUPU project. They also linked the project with the national and local authorities and assisted the project in its national and global advocacy efforts. Meanwhile, since STDM's inception the FIG has supported its development through peer reviews, capacity development, and promotion of the continuum of land rights. Moreover, through its foundation, the FIG has provided grants to support data capture and documentation, and a set of handheld Global Positioning System (GPS) receivers from Trimble, Inc., a contributor to the FIG Foundation. The support facilitated fieldwork implementation.

The *supporters* included the UN-Habitat/GLTN, SDI, MoLHUD, ACTogether, and Municipality of Mbale. The UN-Habitat/GLTN Secretariat, in addition to cofinancing the project, facilitated the planning and consultation workshops and meetings, provided technical support, and facilitated capacity development initiatives. Specifically, the UN-Habitat/GLTN Secretariat led in the customization of STDM for the local context. The SDI was the key implementing partner of the project and provided the necessary political, technical, and administrative support. It mobilized its national network (ACTogether and National Slum Federation) on policy dialogues, advocacy, and project implementation. It also cofinanced the project. The MoLHUD provided the needed political and technical support to the project through the Commission on Urban Development. It enabled the project to be mainstreamed in the TSUPU program and supported its implementation on the ground, particularly by strengthening the linkages with Mbale municipal officials. ACTogether is a leading urban nongovernmental organization (NGO) in Uganda and was the technical arm of project implementation in Uganda. It provided necessary and continuous technical and information support to Mbale Slum Federation and community leaders and members and facilitated liaison and communication between the global, national, and local implementing institutions and stakeholders. The Municipality of Mbale, through its leadership, particularly the mayor and the town clerk, provided enormous technical, coordination, and logistical support. In addition to hosting the slum federation office, it allowed its limited staff and some council members to participate in advocacy, community mobilization and sensitization, and field implementation.

The *implementers* included the Mbale Slum Federation and community leaders and members. The federation took the lead in policy and consultative dialogues with national and Mbale local authorities, community mobilization, sensitization, capacity development, mapping, household interviews, enumeration, and data collection and analysis. They served as the steady bridge between technical and political partners and community members. Meanwhile, at the core of the project implementation are the community leaders and members of the two informal settlements. Most of them are active members of the Mbale Slum Federation. Most of the field enumerators also come from these communities. They participated in the enumeration exercise and were actively engaged in all aspects of the project.

Processes and activities threaded the partnership together. Building from the strengths of Uganda's slum federation, including the tested participatory enumeration processes, the STDM pilot was implemented in the poor settlements of Mission and Bufumbo within the Municipality of Mbale. This included participatory enumeration, a data-gathering process that is to a significant extent jointly designed and conducted by the people who are being surveyed (United Nations Human Settlements Programme/Global Land Tool Network 2010). In summary, the project's key processes and activities implemented included the following.

Planning and consultations: The SDI and the UN Habitat/GLTN Secretariat conducted a series of meetings with authorities (national and municipal), ACTogether, and the slum federation and community members. These yielded an understanding of STDM and its objectives, the finalization of the enumeration questionnaire, the development of an implementation plan, an agreement on the roles and responsibilities, and the identification and mobilization of the needed support and resources.

Mobilization and sensitization processes: The community leaders and members were motivated through an awareness-building process. The Mbale Slum Federation leaders and members spearheaded this process in close collaboration with the municipal officials with technical support from ACTogether. This process generated wide support from the targeted communities and municipal council members. Enumerators were identified from among community members, and they were trained on mapping, data collection, and administration of questionnaires.

Customization of STDM: Following the agreement with the local stakeholders and community members on the enumeration questionnaire and identification of the resources available such as satellite imagery and a handheld GPS, the UN Habitat/GLTN Secretariat proceeded to make adjustments to STDM to customize the system to make it fit the purpose and needs of the local context.

Mapping and structure numbering: With assistance from ACTogether, the slum federation and community members digitized the existing structures from the available satellite imagery and produced initial maps. Using the printed map, assigned enumerators numbered all existing structures and houses in the slum settlements using a unique code. Also, they made use of the handheld GPS to identify available community facilities such as water points, public toilets, dumping grounds, roads, community centers, and so on, as well as newly constructed houses/structures, and update the map accordingly.

Interviews and data collection: As scheduled and communicated with the community, enumeration teams mostly accompanied by local leaders or elders and municipal officials went from house to house to administer the questionnaire. They also collected other information such as supporting documents and photos, with the unique code painted or written on the structure as a background.

Data entry and analysis: The filled-out questionnaires and gathered information, including supporting documents and photos, were entered into the STDM system. This process also included the updating of the initial digital maps. Slum federation leaders and enumerators were trained on how to use STDM to analyze the data and produce reports.

Data validation and continuous updating: As part of quality assurance, the gathered information was validated by community members. This enhanced the acceptability of the information, and all stakeholders appreciated the fact that the turnaround time between the enumeration exercise and the production of results was relatively fast. Enumeration teams and slum federation leaders then updated the information in the system. Some slum federation leaders and members were trained to manage the system and to continue the updating process (Antonio 2011; Antonio et al. 2014).

The overall process is presented in Figure 15.2. The next section discusses the data that became available in this way, as well as their analysis.

Resulting land information, and its use subsequent to the process, was as follows. In the pilot, three data acquisition techniques were combined. First, satellite imagery was used to produce a settlement map wherein structures, houses, roads, water points, and so on were digitized using the STDM plug-in. Second, data from the completed questionnaires were entered into

FIGURE 15.2
Community-driven participatory enumeration process. (From Antonio, D. et al., Addressing the Information Requirements of the Urban Poor—STDM Pilot in Uganda: A Brief, UN-Habitat/Global Land Tool Network, Nairobi, Kenya, 2014.)

an MS Excel spreadsheet for further processing. Once this was done, importing the data into STDM was straightforward. Imported data can be seen, checked, updated, and managed within the STDM data management window. Third, documents, scanned images and texts, photos, and videos can be uploaded into the system. These supporting documents and other file types can link the parties' (individual, household, or groups) tenure status to a specific spatial unit, for example, a structure (as used in the pilot), land, and other properties.

This then allows the use of a powerful tool called Report Builder, which can be used for data analysis and report generation. With this tool, various tenure relationships can be presented, including those with overlaps or that are disputed. The functionality is easy to use and even non-technical users found this functionality very useful, as was mentioned during the pilot by community members, most of them without college education.

With the Report Builder, reports can be automatically generated, including a table of information and/or a specific map showing the selected or queried data or information. Figure 15.3 shows a map highlighting the tenure status of women in Mission settlement as an example.

The STDM plug-in can produce a "certificate" or even a tenure instrument that combines the collected data/information and the generated map. As STDM promotes the continuum of land rights, this "certificate composer" will be very useful once the administrative and/or legal arrangements

FIGURE 15.3
Producing maps.

are in place and authorities (or chiefs for customary tenure arrangements) decide what tenure instrument to provide for poor people. In the pilot area, the communities and authorities have initiated discussions to produce certificates of residency (see Figure 15.4). Community members find it useful as it will open up more development opportunities for them and they believe that such an instrument will be the first step on the ladder toward formalization of their occupation. Community members felt empowered by the spatial information provided by STDM not only because of their use of computers and software but also because the municipality took their demands and needs more seriously. The STDM information empowered people in a way that better enabled them to assert their rights.

Regarding the impact of the pilot, most stakeholders, including slum dwellers, appreciated the added value of the STDM tool in addressing their information requirements. They believe that information is power and that STDM has facilitated that empowerment, and this has strengthened democracy in that community. Moreover, some stakeholders, including government officials, appreciated STDM's potential as a tool for much larger urban development objectives. The impacts or achievements of the project identified include the following:

- STDM was tested and proved to be technically able to address the information requirements of informal settlers. It also helps government authorities in planning purposes.
- Community members are able to use and interact with the STDM software tool and are confident to continuously manage and update the STDM information.

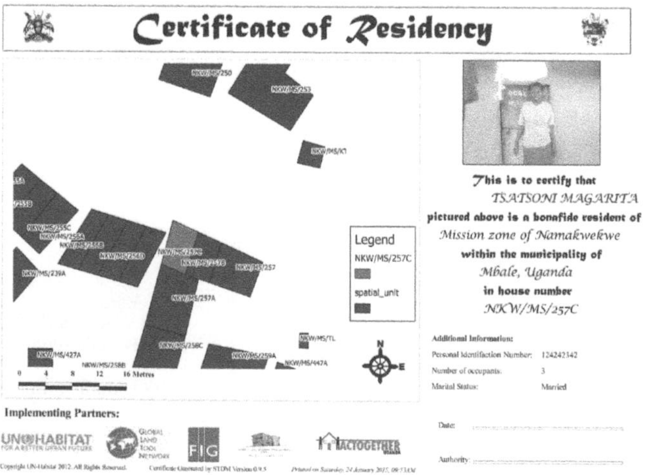

FIGURE 15.4
Generating certificates.

- Data analysis informed communities' plans to pursue priority projects such as roads, lighting, water, and sanitation. Communities are now negotiating with local authorities on possible community development initiatives as informed by data collected using STDM.
- The process provides an opportunity for the authorities and slum communities to initiate dialogues for inclusive planning, access to basic services and infrastructure, and potential tenure security improvement.
- The data generated, for instance, house numbers, are being used as a physical address system that will then enhance slum dwellers' access to other services.
- A regional learning center was established in Kampala, Uganda, for capacity development and future requirements of STDM use and application, particularly in the region.
- Increasing demand to use and apply STDM for several purposes in various situations.
- International acknowledgment of the STDM pilot experience and international recognition of STDM as a tool to promote tenure security, settlement upgrading, and inclusive urban planning.
- Empowerment of communities and community leaders to negotiate with authorities about services, land tenure, and so on and for community land management.

While perceived as being highly successful by most stakeholders, it has faced some specific challenges too, including: (1) the transition in management within ACTogether during project implementation that resulted in some operational delays, (2) challenges encountered in synchronizing timing and interventions of various projects like the TSUPU program, and (3) perceived limited time for sensitization and training according to some enumerators. Valuable lessons were learned during project design and implementation. Based on Antonio et al. (2014), these lessons can be summarized as follows:

- The project is a good model for partnerships. Those between international organizations, national institutions, local authorities, and community stakeholders proved to be the "facilitating" agents of change. The project was well received because all stakeholders were part of the design and implementation.
- STDM proved to be technically sound and simple to use by nonexperts, community members, and community leaders, and it reflects the realities on the ground. Stakeholders appreciated STDM's functionalities in capturing the information on informal settlements using simple technology and the capacities of community members.

They valued the capacity of STDM to generate reports and undertake community-driven analysis in a rapid manner.

- A combination of pro-poor land tools and approaches can be implemented effectively. The project demonstrated that multiple tools and approaches can complement each other such as the slum federation's/SDI's community mobilization approaches and the GLTN's inclusive participatory enumeration process and the STDM tool.
- Ownership by the community of the process is critical for success. The project demonstrated that a people-centered and community-driven approach is vital to the whole process. It is clear that the community members accepted the results as they are the primary owners of the process.
- Capacity development is a catalyst for sustainability. The project highlighted that sustaining the development of a land information management system like STDM is about improving the capacity of users on its use and management.
- Pro-poor solutions have huge potential to impact the lives of the poor if done through strategic partnerships. The project proved that pro-poor land tools like STDM can go a long way particularly if implemented through strategic partnerships. The pilot strengthened the dialogues and engagement of key land stakeholders.
- Use of geo-spatial technologies is no longer the monopoly of experts. The project has established that the use of geo-spatial technologies, like GPS and Geographical Information Systems (GISs), and the use of open and free software packages are no longer the monopoly of educated professionals and, if designed well, even the poor like slum dwellers can use these technologies and benefit from their advantages.
- Enumeration at the city or municipal level may generate more tangible results. It is suggested that if done at the city or municipal level, more outcomes can be expected as the generated information can assist local government authorities in planning for further urban development and rollout of urban services.

Discussion

With the successful implementation of the STDM pilot in Uganda, the demand for its application and implementation has steadily increased, including potential application in other situations. The pilot project is being scaled up to nine municipalities in Uganda in collaboration with the MoLHUD. To date, the slum dwellers federation in Uganda has been able to profile 90 poor settlements in five municipalities. The settlement profiling results in Kampala City revealed that about 72% of the city's population lives in slums and only occupies 12% of the city's land area.

Recently, the slum communities in Mbale municipality were able to gain settlement improvement projects from government authorities capitalizing from STDM-generated data. These projects include construction of inner roads, improvement of drainage systems, and putting up of public sanitation facilities and water points. Also, most municipalities (including Mbale) have hosted the STDM data center within their offices.

STDM is also being implemented in Kenya, Zambia, eastern Caribbean, and Colombia, also with respect to natural resource management (Kenya), agricultural development (Uganda), and post-conflict context (Democratic Republic of Congo). Other GLTN partners and implementing partners at country level are implementing STDM-related projects or programs on settlements. It is clear that STDM is one significant tool that grassroots organizations can use for advocacy and development of objectives, such as inclusive planning, tenure security improvement, and provision of basic services and infrastructure. It is expected that STDM will be implemented in several more countries in the immediate future. It is envisioned that more partnerships will be forged and mobilized in the future, including tapping the support and cooperation of other GLTN partners (i.e., Netherlands' Kadaster and OSGEO), universities and research organizations, IT developers' community, and key individuals. STDM capacity development initiatives have also increased, targeting members of the Young Surveyors Network, which is affiliated to the FIG.

While GLTN partners are confident that STDM can be transformed into a functional land information system for cadastre and land registration systems and other conventional and formal approaches of recording, the objective of its development will continue to focus on providing alternative solutions toward bridging the information divide and targeting the information requirements of the urban and rural poor. STDM development is expected to focus on addressing affordability issues, scalability, simplicity, and participatory and inclusive processes; building on communities' strengths and capacities; and promoting pro-poor and good governance principles (FIG/UN-Habitat/GLTN 2014). With the launching of STDM source codes during the 2014 FIG Congress in Malaysia and setting up of a dedicated website (www.stdm.gltn.net), there is no doubt that its development will be further enhanced and its applications will be further scaled up.

However, it is important to note that STDM is not just an information system or a land tool. It calls for a paradigm shift toward bridging the information divide and toward the recognition of a continuum of land rights approach and maintaining its pro-poor focus. For now, STDM is focused on capturing, recording, analyzing, and generating data and information. STDM and the information that it can generate can be in the hands of the poor with little help and support from authorities, NGOs, and land professionals. Such information and the related processes can be a powerful catalyst for improving the lives of the poor. This will also influence the power balance within the broader land governance context. In case of the STDM pilot in Mbale, the political and administrative authorities from municipal to national levels became really involved. This buy-in was visible, for instance, during the 2012 World Urban Forum (WUF) in Naples, Italy, where a session was organized to showcase the pilot experience. This session was opened by the Uganda minister for housing, land, and urban development, chaired by the commissioner for urban development, and had the mayor and town clerk among its panelists.

A clear challenge in implementing STDM, as learned from the pilot, is how to change the mind-sets of technical people in various institutions, such as land ministries, universities, and the private sector, including land professionals within civil society organizations, to embrace the concepts behind STDM and to work outside the "conventions."

The opportunities for further implementation of STDM in different situations and countries will surely provide more lessons and experience over time. At that time, it will be necessary to take into account the aforementioned considerations with the intention of documenting and adapting to more lessons.

Conclusion

In the last decade, land professionals have developed technical solutions to improve land administration and management systems. With the modernization of the information and communications technologies, such solutions are more powerful, faster, more efficient, and relatively cheaper. Nowadays, organized poor communities and their networks are using advanced technologies and systems with little support from land professionals and are finding them to be vital tools. Indeed, the use of IT systems, remote sensing technologies (i.e., satellite imagery products), Global Navigation Satellite System (GNSS) technologies (i.e., GPS), and GIS systems to create a land information system is no longer the exclusive privilege of the educated elites (United Nations Economic and Social Commission for Asia and the Pacific and the United Nations Human Settlements Programme 2008).

STDM offers related opportunities for land professionals, researchers, grassroots organizations, and government authorities. These opportunities include the empowerment of grassroots communities to develop and manage their own information systems (and data) with less investment of resources and with less reliance on highly paid experts. STDM can build on these strengths and good practices (i.e., enumeration) too. STDM also offers great opportunities for land professionals as they can now extend their services to all, they can now offer people-centered and affordable solutions, and they can also contribute to the further enhancement of the STDM framework. To successfully implement STDM, the different stakeholders mentioned here need to become partners and team up in a partnership, each with its own strengths. With STDM, it is now possible to bridge the information divide and to serve all members of the society and to undertake development interventions such as tenure security for all at scale. Government authorities and decision makers will definitely benefit from its use, recognition, and implementation.

When looking at the five challenges presented in the section "Theoretical Perspective," we can conclude that not only the fourth and fifth challenges have been successfully addressed but also the second one relating to political will has been tackled (as demonstrated by the participation in the WUF session) and was recently demonstrated in the scale-up activities in nine municipalities of Uganda and in other countries. Although less explicitly addressed here, the first and third challenges (focusing on gender and transparency) benefited from the participatory collection and sharing of information. Thus, what started as a pure land information tool grew into a true land governance tool.

STDM will evolve and will be improved over time with the use of more and more partners. It will find other opportunities by being implemented in other contexts (e.g., customary areas) and countries. The consolidated lessons and experience with STDM will inform the strategy for its increasing implementation in the future.

References

Antonio D. 2006. Instituting Good Governance in the Land Administration System—The Philippines' Perspective. In *Land Reform, Land Settlements and Cooperatives*. Rome, Italy: Food and Agricultural Organization of the United Nations.

Antonio D. 2011. Social Tenure Domain Model: Towards Addressing the Information Requirements of Informal Settlements. Paper presented at FIG Working Week 2011, "Bridging the Gap between Cultures," May 18–22, 2011. Marrakech, Morocco.

Antonio D., Gitau J., and Njogu S. 2014. *Addressing the Information Requirements of the Urban Poor – STDM Pilot in Uganda: A Brief*. Nairobi, Kenya: UN-Habitat/Global Land Tool Network.

Augustinus C. 2009. Improving Access to Land and Shelter. In *Innovations in Land Rights, Recognition, Administration and Governance*. Washington: World Bank, GLTN, FIG, and FAO.

Burns A. 2007. *Land Administration Reform: Indicators of Success and Future Challenges*. Agriculture and Rural Development Discussion Paper 37. Washington, DC: World Bank.

Griffith-Charles C. 2010. The application of the social tenure domain model (STDM) to family land in Trinidad and Tobago. *Land Use Policy* 28:514–522.

Deininger K. 2003. *Land Policies for Growth and Poverty Reduction*. Washington, DC: World Bank and Oxford University Press.

Deininger K. and Enemark S. 2010. Land Governance and the Millennium Development Goals. In *Innovations in Land Rights, Recognition, Administration and Governance*. Washington: World Bank, GLTN, FIG, and FAO.

Enemark S., Bell K.C., Lemmen C., and McLaren R. 2014. *Fit-For-Purpose Land Administration*. FIG Publication No. 60. Copenhagen, Denmark: FIG/World Bank.

FIG/UN-Habitat/GLTN. 2014. *A Review of Social Tenure Domain Model (STDM) Phase 2–Summary Report*. Copenhagen, Denmark: International Federation of Surveyors (FIG).

Food and Agriculture Organization of the United Nations. 2007. Good Governance in Land Tenure and Administration, FAO Land Tenure Series 9, Rome, Italy.

Henssen, J.L.G. 1995. Basic principles of the main cadastral systems in the world. In: *Modern Cadastres and Cadastral Innovations*, Proceedings of the One Day Seminar in Delft, May 16, 1995. FIG Commission 7 and University of Melbourne, pp. 5–12.

Lemmen C. 2010. *The Social Tenure Domain Model—A Pro-Poor Land Tool*. Copenhagen, Denmark: Global Land Tool Network, United Nations Human Settlements Programme, and International Federation of Surveyors.

Lemmen C. 2012. *A Domain Model for Land Administration*. Delft, The Netherlands: Sieca Repro BV.

Lemmen C., Augustinus C., Haile S., and van Oosterom P. 2009. *The Social Tenure Domain Model—A Pro-Poor Land Rights Recording System*. The Netherlands: GIM International.

Magel H. 2006. Promoting Land Administration and Good Governance. A keynote address to the 5th FIG Regional Conference, Accra, Ghana.

Palmer D., Fricska S., and Wehrmann B. 2009. *Towards Improved Land Governance*. Land Tenure Working Paper 11. Rome, Italy: Food and Agriculture Organization of the United Nations and the United Nations Human Settlements Programme.

Transparency International. 2009. *Global Corruption Barometer 2009*. Berlin, Germany.

Transparency International-India. 2005. *India Corruption Study 2005*. New Delhi, India.

United Nations Economic and Social Commission for Asia and the Pacific and the United Nations Human Settlements Programme. 2008. *Housing the Poor in Asian Cities—Land: A Crucial Element in Housing the Urban Poor*. Bangkok, Thailand.

United Nations Human Settlements Programme. 2008a. *State of the World's Cities 2010/2011—Cities for All: Bridging the Urban Divide*. GLTN: Nairobi, Kenya.

United Nations Human Settlements Programme. 2008b. *Secure Land Rights for All*. GLTN: Nairobi, Kenya.

United Nations Human Settlements Programme/Global Land Tool Network. 2010. *Count Me in—Surveying for Tenure Security and Urban Land Management*. Nairobi, Kenya.

United Nations Human Settlements Programme/Global Land Tool Network. 2012. *Handling Land*. Nairobi, Kenya.

van Oosterom P.J.M., Lemmen C.H.J., Ingvarsson T., van der Molen P., Ploeger H.D., Quak W. et al. 2006. The core cadastral domain model. *Computers, Environment and Urban Systems* 30 (5):627–660.

Williamson I., Enemark S., Wallace J. and Rajabifard A. 2010. Land Administration for Sustainable Development, USA, ESRI Press Academic.

Zevenbergen J., Augustinus C., Antonio D., and Bennett R. 2013. Pro-poor land administration: Principles for recording the land rights of the underrepresented. *Land Use Policy* 31:595–604.

Section V

Looking Ahead

16

Future Directions in Responsible Land Administration

Jaap Zevenbergen, Walter T. de Vries, and Rohan M. Bennett

CONTENTS

Embracing Harmonious Dynamism

Fundamentally, the term "administration" suggests bureaucratic, controlled, and steady, if not slow, paces of change. However, the previous chapters have shown that the relations between people and land that land administration attempts to capture are the very opposite. At all levels of abstraction, land administration can be seen as being multifaceted, crosscutting, interdisciplinary, and above all dynamic. Dynamism in land administration is currently visible in the social and political recognition—or negotiations on recognition—of land tenure typologies (Simbizi et al. 2014; van Leeuwen 2014). The developments in geo-ICT create their own dynamics. They offer the opportunity for previously unforeseen methods of land data capture, visualization, and sharing (Uitermark et al. 2010). Geo-ICT disturbs more than the technical elements of land administration systems: organizational and political contingencies are placed in flux when technology selections are made (Kurniawan and de Vries 2015; Sui 2014).

Dynamism in land administration is most prominently viewed in large-scale land tenure regularization programs, usually at the national level. Formal recognition of land rights changes the status of people, land, and the relationship between them. Fit-for-purpose approaches to data collection and management change the core characteristics of land administration: systems become flexible, inclusive, participatory, affordable, reliable, attainable, and upgradeable (Enemark et al. 2014). After the intervention, the perceptions of landholders change: they might invest in the land and transfer it to other people, ultimately changing land use and land value. The changes must be monitored and evaluated, particularly in the contemporary era where the accountability of donor agencies, and all parties involved in the programs, is heightened. Measuring the interventions is no trivial task: isolating meaningful dependent and independent variables is an ongoing challenge. Socio-technical approaches are needed as are skilled personnel to implement them. This suggests substantial changes to social capacity, embodied in scaled capacity-building programs: to reap the rewards of well-designed interventions, integrated capacity development activities are needed at individual, cross-organizational, and societal levels. When all the aforementioned changes coalesce in a harmonious fashion, "responsible land administration" appears to be more readily achievable.

Synthesizing Inspired Advances

This book shows that we are advancing toward responsible land administration. There is a drive to document people to land relationships, regardless of the type of tenures the records represent. With this mind-set, new inspirations are driving new designs, and new evaluation approaches enable the impacts to be measured. Each chapter of this book contains one part of this story of change: five levels of change are observed as crosscutting themes in this book.

In relation to *changes in people to land relationships*, Chapters 5 and 14 show the critical role that land plays in post-conflict situations and how land administration changes during post-conflict periods. Similarly, Chapters 2, 3, 6, and 11 include conceptual insights on how alternative people to land relationships can be modeled to accommodate the poorest in a society, food security, crowds, and nomads. Chapter 15 discusses elements of the Social Tenure Domain Model (STDM), a basis for an alternative way for recognizing and recording people to land relations. Land is not only used or possessed by single persons for static parcels: dynamic models allow nomadic land uses to be documented as well (Chapter 11). In other words, the new types of relations conceptualized within responsible land administration are increasingly becoming an indispensable ingredient for solutions in post-conflict governance, food security provision, and poverty alleviation. In addition, "the state" is perceived to no longer monopolize the acceptance of people to land relationships: crowds and clouds form part of an emerging alternative.

Regarding *changes in conceptual understandings and technological possibilities*, innovative designs, such as the point cadastre in Chapter 7 and the use of simple tools like the digital pen in Chapter 8, showed us that one can represent an interest in land with a mere point, as a first step. Increasing information-processing power, alongside rapidly developing software alternatives, provided further opportunities for unconventional record and workflow management. In this respect, Chapter 9 demonstrated a design for ensuring more rapid recording of changes in people to land relationships in a land information system: to ensure that the system is kept up-to-date, a mix of pragmatism and multiple options for transacting parties was proposed. Meanwhile, Chapter 10 applied the new technologies to simulate and stimulate improving land uses, via land consolidation.

In relation to *land use changes*, Chapters 4 and 12 focus on urban land use changes and systems to manage and administer these changes better: as human populations rapidly urbanize, land becomes a crucial issue for good urban governance. In contrast, Chapters 10, 11, and 13 focus on examples of changes and challenges in rural areas. Demonstrating synergies between the two contexts, underlying techniques studied in rural areas, such as land consolidation, also has applications in urban settings.

In relation to *measuring the changes*, good impact measurement informs intervention management and in some cases redesign. Chapters 12 and 13 specifically address alternative evaluation methods. Combining the perceptions of people with remotely sensed land use changes helped to visualize and understand impacts. At first glance, it seems there are no winners following one intervention: users are pushed from the land and even from having access for the greater good of society (i.e., nature preservation). A better analysis of each stakeholder's needs helps to find a win–win solution. Similarly, Chapters 5 and 14, related to post-conflict governance and protracted displacement, show how rapid societal changes impact land administration, and vice versa. Although the methodologies applied in these chapters focus on violent conflicts, one could also apply them in contexts where natural disasters have occurred. In both cases, they result in multiple layers of claimants from different periods of time, each of them requiring recognition, administration, and a certain level of legitimacy. Different solutions for giving land to each user were trialed in post-conflict Rwanda: peace was kept, but it also influenced the later land tenure regularization intervention.

In relation to *change agents*, various chapters indirectly identify new types of (responsible) land administrators: these actors look beyond their traditional professional and academic boundaries. For example, the point cadastre land administrators of Chapter 7 are no longer equal to conventional land surveyors or adjudicators. The IT and workflow managers addressed in Chapter 9 are neither pure computer scientists nor pure topographic or cadastral managers. The crowds addressed in Chapter 6 and the actor network analysts are completely, or relatively, new actors in the land sector, a field that has tended to be heavily regulated by licenses and accreditation instruments. Information is another change agent: if "information is power," documenting people to land relationships not only influences tenure security at the individual landholder level but also empowers different groups of landholders with respect to local and national governments. If land interventions are executed well, they not only give us "responsible land administration" but also make contributions to better land governance and toward shared prosperity.

Further Responsible Pursuits

Beyond the advances illustrated in the book, further work awaits the domain of responsible land administration.

Regarding *changes in people to land relations*, there are a number of new opportunities. First, rapidly changing social relations, embodied by population increases, especially in urban areas, mean that further alternative pro-poor system designs will be needed. In particular, peri-urban areas need attention. These zones are often home to the poorest in society but usually fall into administrative voids between formal and customary systems (Zevenbergen et al. 2013). Second, the conceptual thinking of the "continuum of land rights" (UN-Habitat 2008), developed to illustrate the diverse ways that individuals and groups exercise rights to land, has proved useful to further the conceptual discourse on land rights; however, it remains a debated concept under conceptual development (Whittal 2014). Further development and testing of the continuum concept should proceed, and this will affect the context of responsibility in which land administrators need to operate. Third, understandings of post-conflict contexts and the role of land administration in such contexts are improving; however, there is still a need to better conceptualize and cross-compare cases. Some of these conceptual studies have already begun (Hollingsworth 2014). Fourth, "neocadastres," a concept first introduced by de Vries et al. (2014), refer to the new institutional relations that emerge as a result of the interaction between the newly emerging technologies (in the broad sense) and the conventional institutional settings of land agencies. The main question hereby is to which extent, and if so in which way, the conventional cadastral norms are altered in view of persuasive norms and associated practices of the new technologies. Given that institutional norms tend to be rather rigid, and largely rooted in historical practices, it is not easy to alter them, even though altering may be more appropriate given the changes in organizational or technical context. Finally, the era of "smart land administration" beckons. Analogous to smart cities, smart government, and smart mobility, smart land administration refers to the kind of processes of land administration that uses social technologies, volunteered geographic information, and crowdsourcing in combination with technical drivers of intelligent information systems and big, linked, and open data. Ultimately, this will supposedly drive "smarter" solutions for land-related challenges.

The question, however, remains whether these solutions, which tend to have a technological bias, are necessarily more responsible.

In relation to *changes in conceptual understanding and technological possibilities*, several agitations are apparent. First, STDM has been the basis for an open source software package, which is still under development and testing. It is not the only open source package that is currently available, which indicates a certain trend in technology development and uptake. Land administration can tap into this trend and use it to its advantage. All this is currently embedded in the discourse toward the "Global Cadastre" (McDougall et al. 2013). Second, unmanned aerial vehicles (UAVs) are increasingly available for citizens and nongovernmental organizations (NGOs), among others. Various tests are conducted to verify their application in the land administration domain, for example, mapping or monitoring informal settlement areas or rapid assessments after disasters. The application of such technologies can connect with or extend to earlier experiments of point cadastres, for example. Third, on the basis of terrestrial digital photogrammetry there are new ways of creating three-dimensional (3D) models. Mobile phone-based data collection tools, in combination with smart software, can be used to create 3D models—including 3D cadastral objects—in contexts where conventional methods do not allow this. This is an alternative for large-scale mapping, with relatively low-cost tools, and especially suitable for areas where there are no, or few, two-dimensional (2D) or 3D cadastral data. Fourth, 3D, four-dimensional (4D), and five-dimensional (5D) system modeling (Van Oosterom and Stoter 2010) is on the horizon, if not already here: including the height dimension is increasingly important in administrating the boundaries of urban properties. Added to this is the time dimension of both spatial and legal attributes. Combined, they create a new kind of registration system, for which the implications are still not evident (Kitsakis and Dimopoulou 2014). Five, "Open Cadastres" are also emerging. This term primarily refers to the uptake of social media and open source technologies for the purpose of administration of land rights (Basiouka and Potsiou 2014; Laarakker and de Vries 2011). Six, "Green Cadastres" are also hypothesized (Bennett et al. 2012). These are cadastres that record the property interests of the natural environment. Climate change responses and unbundling of the conventional land parcel drive the creation of these new interests: a reconceptualization of the characteristics and spatial nature of real property is needed.

In relation to *land use changes*, driving forces and actors will interact and potentially produce significant changes: land administration systems will be called on to support responsible governance of these changes. First, as mentioned, in urban and peri-urban areas land readjustment programs will be part of improved urban governance agendas. These activities will be aided by information about the landholdings, potentially found in a pro-poor cadastral system. Second, the increased prominence of women's access to land in international and national policy agendas will see changes in recognized land users, and land use: land administration systems will need to support, reflect, and embed these changes. Third, responses to climate change might demand wholesale changes to land use in some areas while promoting continuous sustainable practices (e.g., nomadic pastoralism) in others: land information will support change management but will also need collection in places where it currently does not exist. Fourth, in the related area of food security responses will call on the empowerment of smallholders, and their increased production output: new sustainable and inclusive land consolidation approaches, supported by land administration systems, will be needed in this respect.

In relation to measuring the changes, one overarching initiative needs attention. Evidence of the existence, and effectiveness, of pro-poor approaches needs to be assessed, especially with respect to the verified achievement of Millennium Development Goals (MDGs) and their post-2015 development agenda replacement. For the MDGs, it often remains unclear to which baselines conventional evaluation schemes are measuring, and to which extent, or under which conditions land administration and poverty alleviation can be truly connected. Indeed, this appears to be one of the key challenges for responsible land administrators, namely, developing and testing alternative methods of measurement, evaluation, and attribution. Evaluation tools for understanding the role land administration plays, and can play, in supporting a whole range of objectives are needed, including food security, climate change, gender equality, anti-land-grabbing activities, positive land consolidations, equitable land readjustments, and alternative urban tenure security models, to name a few.

In relation to change agents, a few points can be made. First, a new type of land administrator is emerging. He or she must realize the broad impacts of land interventions on all aspects of sustainability (social, economic, and environmental), the political economy between those depending on cheap land access and those playing with landed property as a near abstract asset, and the finite nature of (useable) land on an increasingly populated planet. He or she must understand the different dogmatic positions of idealized types of people to land relations and their documentation, including the fact that the day-to-day reality is never black or white but only has contextualized shades of grey. Second, new ways of designing and conducting capacity development assessments are needed. Understanding that capacity is very much related to local context and the stakeholders' frames and needs embedded in this context implies that capacity development is no longer a static exercise of determining a fixed gap in eternally required skills and knowledge. It implies that one has to continuously assess how frames and needs are changing in time, place, and political and organizational contexts and how provision of skills, experience, and knowledge can cater to these given a limited set of resources. With respect to the change agents, part of the responsibilities lies with institutes that train land professionals or, better, that develop capacity in the land sector. Universities with land-related programs play an important role in this, as they are usually at the top of the "training food chain." Too much focus on disciplinary knowledge and skills can even be counterproductive and lead to suboptimal choices in land interventions. Land professionals need as much development of their attitudes as their technical skillset.

The ITC Faculty of the University of Twente, Enschede, The Netherlands, aims to contribute to the aforementioned type of capacity development via change agents, who can understand and practice in the ever dynamic land administration domain. Through this book, we feel that the chapters based on MSc and PhD research done within the ITC Faculty show that we are contributing to this aim. We also realize that we are not alone in this endeavor: we cannot cover all relevant angles and dimensions. We are pleased to be part of a community of networks of like-minded people, both individuals and organizations. Currently, the Global Land Tool Network hosted by the UN Habitat, with more than 65 partners and more than one-third of them within the "training and research cluster," is a key vehicle for this. We are therefore also pleased to contribute to the thinking toward a core land curriculum that encompasses some of the aforementioned ideas and philosophies. Responsible land administration is advancing; but further work lies ahead, and this can only be done through a worldwide effort led by change agents, one that rattles the foundations of conventional, disciplinary notions of land administration components.

Final Remarks

Responsible land administration is especially developing in the context of the Anthropocene—an urban age where global challenges of rapid and massive scale urbanization and migration coincide with major conflicts relating to land, food security, water, infrastructure, and other resources. Therefore, land administration needs to scale up efforts and integrate with other domains, incorporate new axioms, and seek out new paradigms and research questions. Such research questions are mainly socio-technical and institutional in nature, creatively combining globally available technologies with a clear understanding of a legal, organizational, and governance context. In this way, land administration can further develop into a new type of scientific discipline, one that can support the derivation of contemporary fit-for-purpose and responsible solutions.

References

Basiouka, S., and C. Potsiou. 2014. "The volunteered geographic information in cadastre: perspectives and citizens' motivations over potential participation in mapping." *GeoJournal* 79 (3):343–355.

Bennett, R.M., P. van der Molen, and J.A. Zevenbergen. 2012. "Fitted, green, and volunteered, legal and survey complexities of future boundary systems." *Geomatica* 66 (3):181–193.

de Vries, W.T., R.M. Bennett, and J.A. Zevenbergen. 2014. "Neo-cadastres: innovative solution for land users without state based land rights, or just reflections of institutional isomorphism?" *Survey Review* 47 (342):220–229. doi: doi:10.1179/1752270614Y.0000000103.

Enemark, S., K.C. Bell, C.H.J. Lemmen, and R. McLaren. 2014. *Fit-for-Purpose Land Administration: Open Access e-Book*. Vol. 60, FIG publication. Copenhagen, Denmark: International Federation of Surveyors (FIG).

Hollingsworth, C. 2014. "A Framework for Assessing Security of Tenure in Post-Conflict Contexts." MSc Geoinformation Science and Earth Observation, University of Twente, Enschede, The Netherlands.

Kitsakis, D., and E. Dimopoulou. 2014. "3D cadastres: legal approaches and necessary reforms." *Survey Review* 46 (338):322–332. doi: doi:10.1179/1752270614Y.0000000119.

Kurniawan, M., and W.T de Vries. 2015. "Impact of geoICT applications on speed and uptake of local government services." *Local Government Studies* 41 (1):119–136. doi: 10.1080/03003930.2014.937001.

Laarakker, P., and W.T. de Vries. 2011. *www.Opencadastre.org—Exploring Potential Avenues and Concerns*. FIG Working Week. Bridging the Gap between Cultures, Marrakech, Morocco.

McDougall, K., R.M. Bennett, and P. van der Molen. 2013. "The global cadastre: improving transparency in international property markets." *GIM International* 27 (7):30–34.

Simbizi, M.C.D., R.M. Bennett, and J. Zevenbergen. 2014. "Land tenure security: revisiting and refining the concept for Sub-Saharan Africa's rural poor." *Land Use Policy* 36 (0):231–238. doi: http://dx.doi.org/10.1016/j.landusepol.2013.08.006.

Sui, D. 2014. "Opportunities and impediments for open GIS." *Transactions in GIS* 18 (1):1–24. doi: 10.1111/tgis.12075.

Uitermark, H., P. Van Oosterom, J. Zevenbergen, and C. Lemmen. 2010. *From LADM/STDM to a Spatially Enabled Society: A Vision for 2025*. The World Bank Annual Bank Conference on Land Policy and Administration, Washington DC. April 26–27, 2010.

UN-Habitat. 2008. *Secure Land Rights for All*. Nairobi, Kenya: GLTN.

van Leeuwen, M. 2014. "Renegotiating customary tenure reform—land governance reform and tenure security in Uganda." *Land Use Policy* 39 (0):292–300. doi: http://dx.doi.org/10.1016/j.landusepol.2014.02.007.

Van Oosterom, P., and J. Stoter. 2010. "5D data modelling: full integration of 2D/3D space, time and scale dimensions." In Fabrikant S.I., Reichenbacher T., van Kreveld M., Schlieder C. (eds.) *Geographic Information Science*, pp. 310–324. Berlin, Germany: Springer-Verlag.

Whittal, J. 2014. "A new conceptual model for the continuum of land rights." *South African Journal of Geomatics* 3 (1):13–32.

Zevenbergen, J., C. Augustinus, D. Antonio, and R. Bennett. 2013. "Pro-poor land administration: principles for recording the land rights of the underrepresented." *Land Use Policy* 31:595–604. doi: 10.1016/j.landusepol.2012.09.005.

Index

D

G

H

I

M

Milton Keynes UK
Ingram Content Group UK Ltd.
UKHW020314111024
449327UK00040B/1089